SpringerBriefs in Electrical and Computer Engineering

For further volumes:
http://www.springer.com/series/10059

Rose Qingyang Hu • Yi Qian

Resource Management for Heterogeneous Networks in LTE Systems

Springer

Rose Qingyang Hu
Department of Electrical
    and Computer Engineering
Utah State University
Logan, UT, USA

Yi Qian
Department of Computer
    and Electronics Engineering
University of Nebraska-Lincoln
Omaha, NE, USA

ISSN 2191-8112 ISSN 2191-8120 (electronic)
ISBN 978-1-4939-0371-9 ISBN 978-1-4939-0372-6 (eBook)
DOI 10.1007/978-1-4939-0372-6
Springer New York Heidelberg Dordrecht London

Library of Congress Control Number: 2014933579

Printed on acid-free paper

Springer is part of Springer Science+Business Media (www.springer.com)

To our respective families.

Preface

As a key technology in 4G-LTE, heterogeneous networks effectively extend the coverage and capacity of wireless networks by deploying multiple low power small base stations on top of the conventional macro base stations. The deployed small nodes differ in transmission power and processing capabilities, leading to new challenges in mobile association, interference management, and radio resource management. In this book, we consider downlink communications in a heterogeneous cellular network with high transmit power macro evolved Node Bs and low transmit power small evolved Node Bs. We provide an in-depth look on the key issues that could affect the performance of heterogeneous networks and present schemes that can effectively tackle these issues. In particular, we discuss the issue of unbalanced traffic load among the macro evolved Node Bs and small evolved Node Bs caused by the transmit power disparity and present a load-balancing based mobile association scheme to balance the traffic load among the macro evolved Node Bs and small evolved Node Bs. We explore the issue of high intra-cell interference received by the user equipment associated with the small evolved Node Bs from the high power macro evolved Node Bs and introduce a fractional frequency reuse scheme with proper power control to help reduce interference at user equipment that are the most vulnerable to such intra-cell interference. We investigate radio resource allocation issues for heterogeneous networks with intracell cooperation and propose a resource allocation framework that could achieve the maximum capacity with proportional fairness among user equipment. For each of the investigated issues and presented solutions, we also present numerical results to demonstrate the effectiveness of the proposed solutions in tackling the problems and improving network performance.

Logan, USA   Rose Qingyang Hu
Omaha, USA   Yi Qian

Acknowledgements

First of all, we would like to thank all the students and post-doctoral researchers in both our research groups at Utah State University and University of Nebraska-Lincoln, who have worked on the related research topics and contributed to the research activities as well as helped to draw all the figures in this book. Secondly, we would like to thank all the reviewers for their dedicated time in reviewing this manuscript, and for their valuable comments and suggestions for improving the quality of this book. Finally, we appreciate the advice and support of the staff members from Springer, for putting this book together.

This work was supported by US National Science Foundation (NSF) grants ECCS-1308006 and ECCS-1307580.

Contents

Chapter 1
Introduction

Mobile communications, which enable anytime and anywhere ubiquitous connectivity, have been an integrated part of our daily life. With the widespread adoption of smart mobile devices such as smart phones, tablets and ultra-portable laptops and the subsequent explosive expansion of bandwidth-hungry mobile applications, wireless communication traffic continues to grow rapidly. As voice traffic grows at a steady rate, the major traffic explosion comes from data communications. According to Cisco's Visual Networking Index, data traffic has been more than doubled in both 2011 and 2012. Projection through 2018 expects a compound annual growth rate (CAGR) of around 50 % [1]. To sustain such a traffic growth and in the meantime to improve user experience and network coverage, continuous innovations on wireless data communication technologies are required.

In the process of technology innovation, the 3rd Generation Partnership Project (3GPP) plays a leading role. Initiated in 1998, 3GPP family of technologies have evolved from GSM-EDGE, to UMTS-HSPA-HSPA+, to LTE and LTE-Advanced. HSPA/HSPA+ has been widely deployed nowadays providing theoretical data rate of up to 168 Mbps in the downlink (DL) and 22 Mbps in the uplink (UL). HSPA/HSPA+ enables diverse data applications and enhanced user experience, which further foster the rapid advancement of information and communication industry. According to 4G Americas, as of September 2012, 476 commercial HSPA networks have been deployed in 181 countries worldwide. As 3GPP continues to work towards further enhancements on HSPA/HSPA+, Long Term Evolution (LTE) system was introduced in 3GPP Specifications Release 8 (Rel-8) with the aim to provide an even higher data rate and better user experience, and to eventually fulfill the International Mobile Telecommunication-Advanced (IMT-Advanced) requirements issued by the International Telecommunication Union Radio-telecommunication Sector (ITU-R). An LTE system consists of a flat IP-based evolved packet core (EPC) and an OFDMA-based radio access network (RAN), also known as Evolved Universal Terrestrial Radio Access Network (E-UTRAN). 3GPP Rel-8 (forzen in December 2008) together with a follow-up Rel-9 (forzen in December 2009) defined the specifications for LTE. As 3GPP evolves towards Rel-10, a bunch of new

R.Q. Hu and Y. Qian, *Resource Management for Heterogeneous Networks in LTE Systems*, SpringerBriefs in Electrical and Computer Engineering, DOI 10.1007/978-1-4939-0372-6_1, © The Author(s) 2014

technologies were introduced such as carrier aggregation (CA), multiple-antenna enhancement, self-organizing network (SON), multimedia broadcast/multicast services (MBMS) and heterogeneous networks (HetNet), etc. Following the frozen of Rel-10 in March 2011, work on Rel-11 started. 3GPP Rel-11 focuses on enhancement on technologies introduced in Rel-10, such as enhanced CA, HetNet, MBMS and SON. Co-ordinated Multi-Point (CoMP) transmission and reception are also introduced in Rel-11. By December 2012, the majority of Rel-11 work items had been completed. 3GPP specifications Rel-10 and Rel-11 constituted to LTE-Advanced. In October 2010, ITU-R Working Party 5D agreed that LTE-Advanced had met all the requirements of IMT-Advanced. LTE-Advanced has been incorporated as one of the two radio interfaces for IMT-Advanced (the other one is WiMAX).

As 3GPP Rel-11 approaches to its completion, planning for Rel-12 started in June 2012. The goal for 3GPP Rel-12 and beyond is to meet the projection of $1,000\times$ capacity increase by 2020. As air interface approaches its theoretical capacity limit and new spectrum is difficult and costly to obtain, capacity increase is expected to mainly come from network architecture improvement. Heterogeneous network architecture with a multi-tier multi-communication network deployment is understood to be a promising direction for evolution. Toward this direction, small cell enhancement, device-to-device (D2D) communications and LTE-WLAN interworking become the key study items for 3GPP Rel-12. Working beyond Rel-12, there will be further progress in deploying heterogeneous network architecture.

Definition for heterogeneous networks is quite loose and diverse. Some people consider a network with the overlay of macro cells and small cells (micro, pico, femto) of the same air interface as a heterogeneous network. Others consider cellular network plus WLAN as the main use case. To the authors, heterogeneous network is an integration of diverse technologies and network architectures in achieving high spectrum/energy efficiency and quality-of-service (QoS). A general heterogeneous network consists of multiple tiers of networks of different cell sizes/footprints and/or of multiple radio access technologies. In this book, it will focus on a two-tier heterogeneous network model with a single radio access technology based on LTE-Advanced.

The two-tier heterogeneous network is featured by a joint deployment of macro cells of wide coverage and high transmit power macro evolved Node Bs (eNBs) and small cells of limited coverage and low transmit power small eNBs. The small cells could be deployed within the coverage of macro cells for data rate enhancement or out of the coverage of macro cell for coverage extension. The co-existence of high transmit power macro eNBs (typically 43–46 dBm) and low transmit power small eNBs (typically 24–37 dBm) arises new problems in mobile association, radio resource management, and mobility management. In this book, each of the above mentioned problems associated with heterogeneous networks will be discussed and solutions will be proposed in addressing these problems.

Conventional way of doing mobile association is based on a best-power based approach where each user equipment (UE) associates with the eNB that the UE receives the highest power from. However, in heterogeneous networks, due to the disparity in transmit power of the macro and small eNBs, if best-power based mobile association is again used, most of the UEs will associate with the macro eNBs leaving the small eNBs largely under-utilized. This would undermine the ability of small cells in traffic-offloading and data rate enhancement. To deal with this problem, it was proposed to use a range-expansion based mobile association where a bias is used to compensate the power difference between macro and small eNBs so that more mobiles can be associated with small eNBs [2]. However, a proper bias value remains to be determined. Since then, more mobile association schemes have been studied in heterogeneous networks. In [3], a new mobile association scheme is proposed in a heterogeneous wireless network, where a mixed deployment of macro eNBs and small relay nodes can lead to uplink and downlink imbalance of a mobile. The proposed scheme will allow the mobiles always connect to the best access node(s) on both uplink and downlink. In [4], a jointly optimal mobile association and load balancing framework that aims to maximize the relay network capacity is presented. In addition, a heuristic algorithm that enables a practical implementation is described. In [5], the main challenges for mobile association and load balancing in a heterogeneous network with relay nodes are discussed and addressed. An optimal framework that aims to maximize system capacity is presented. It considers both the relay backhaul resource usage and wireless access link resource availability. A heuristic algorithm that enables the practical implementation is proposed and evaluated. In [6], a load-balancing based mobile association is proposed where the mobile association problem is formulated as an optimization problem with the objective of minimizing a weighted total resource consumption at the macro and small eNBs for a given number of UEs. It shows that the proposed load-balancing based mobile association can achieve an efficient usage of the radio resources at both the macro and small cell eNBs and effectively improve network throughput. In [7], a new mobile association framework is proposed for the heterogeneous wireless networks with intra-cell node cooperations. The proposed framework aims to maximize the network capacity and balance the traffic load among the network nodes. Furthermore, a pricing mechanism is introduced to enable the distributed implementation of the proposed scheme. In [8], a new mobile association scheme is further studied that jointly maximizes downlink system capacity and minimizes the mobile station (MS) uplink transmitting power. Simulation results show that a significant performance gain on the defined objective function is achieved.

Another issue in heterogeneous networks is radio resource management. As in homogeneous networks, it needs to decide the frequency reuse among the macro eNBs and the resources allocated to the UEs. Moreover, in heterogeneous networks, the frequency reuse scheme between the macro and small cell eNBs also needs to be decided. For relay with in-band backhaul, the resource partition between relay access link and its backhaul needs to be carefully planned. It needs to make sure that the backhual link does not consume excessive radio resources or becomes the bottleneck limiting the full utilization of the relay resources. In [9] and [10], several resource coordination schemes have been studied in time, frequency, power dimensions in an heterogeneous LTE relay network. Furthermore, an optimal fractional frequency reuse and power control scheme has been proposed that can effectively coordinate the interference among high power and low power nodes. The scheme can be optimized to maximize the total long term log-scale throughput among all the UEs.

A downlink intra-cell cooperative transmission in the heterogeneous networks is studied and an optimal cooperation scheme to achieve both throughput maximization and user fairness is developed in [11]. The scheme is optimized by selecting the best SINR threshold to form intra-cell cooperation. The optimization is based on long term time averaged system information and only needs to be updated pseudo-dynamically. An optimal intra-cell coordinated multipoint processing resource allocation scheme in a wireless heterogeneous network is presented and a precoding method in the physical layer to reduce the inter-user interference is explored in [12]. The proposed scheme can improve the network capacity and coverage considerably. Radio resource allocation for heterogeneous networks with cooperative relays is investigated in [13], where the relay nodes with in-band backhaul act as small eNBs and are able to serve UEs either independently or cooperatively with the macro eNBs. A radio resource-allocation framework is proposed and a resource-allocation strategy is derived that is asymptotically optimal on the proportional fairness metric. The derived resource-allocation scheme gives insights on the optimal radio resource allocation for the heterogeneous networks with cooperative relays using in-band backhauls.

In this book, a unified HetNet model in a general LTE system is studied with high transmit power macro eNBs and low transmit power small eNBs or relay nodes. Radio resource management schemes in such a system are explored. The rest of the book is organized as follows. In Chap. 2, the details of the general HetNet network model are provided and the background information is introduced for the key resource allocation techniques applied in HetNet such as mobile association, frequency reuse and interference management, and cooperative multi-point transmission. In Chap. 3, mobile association schemes for HetNets are presented. In Chap. 4, inter-cell interference coordination schemes with fractional frequency reuse for the HetNets are studied. In Chap. 5, radio resource allocation schemes in HetNets are further investigated. In all these three chapters, resource allocation schemes are presented for enhancing network spectrum efficiency for HetNets in LTE systems.

References

1. Cisco, "Cisco visual network index: Global mobile traffic forecast update," 2012.
2. Qualcomm Europe, "Range expansion for efficient support of heterogeneous networks," TSG-RAN WG1 #54bis R1-083813, Sep. 2008.
3. R. Q. Hu, Y. Yu, Z. Cai, J. E. Womack, Y. Song, "Mobile Association in a Heterogeneous Network," in *Proc. of IEEE ICC 2010*, Cape Town, South Africa, May 2010.
4. Y. Yu, R. Q. Hu, C. Bontu, Z. Cai, "Mobile Association and Load balancing in a Cooperative Relay Enabled Cellular Network," *IEEE Communications Magazines*, 49(5), pp. 83–89, May 2011.

5. Y. Yu, R. Q. Hu, Z. Cai, "Optimal Load Balancing and Its Heuristic Implementation in a Heterogeneous Relay Network," in *Proc. of IEEE GLOBECOM 2011*, Houston, Texas, Dec. 2011.
6. Q. C. Li, R. Q. Hu, G. Wu and Y. Qian, "On the optimal mobile association in heterogeneous wireless relay networks," in *Proc. of IEEE INFOCOM 2012*, Orlando, FL, Mar. 2012.
7. Q. C. Li, Y. Xu, R. Q. Hu, G. Wu, "Pricing-based distributed mobile association for heterogeneous networks with cooperative relays," in *Proc. of IEEE ICC 2012*, Ottawa, Canada, Jun. 2012.
8. R. Chen, R. Q. Hu, "Joint uplink and downlink optimal mobile association in a wireless heterogeneous network," in *Proc. of IEEE GLOBECOM 2012*, Anaheim, CA, USA, Dec. 2012
9. R. Q. Hu, Y. Qian, and W. Li, "On the downlink time, frequency and power coordination in an LTE relay network," in *Proc. of IEEE GLOBECOM 2011*, Houston, Texas, Dec. 2011.
10. Q. C. Li, R. Q. Hu, Y. Xu, and Y. Qian, "Optimal Fractional Frequency Reuse and Power Control in the Heterogeneous Wireless Networks," *IEEE Transactions on Wireless Communications*, Vol. 12, No. 6, pp. 2658–2668, Jun. 2013.
11. Y. Xu, R. Q. Hu, "Optimal Intra-cell Cooperation in the Heterogeneous Relay Network," in *Proc. of IEEE GLOBECOM 2012*, Anaheim, CA, Dec. 2012.
12. Y. Xu, R. Q. Hu, Q. C. Li, Y. Qian, "Optimal Intra-cell Cooperation with Precoding in the Wireless Heterogeneous Networks," in *Proc. of IEEE WCNC 2013*, Shanghai, China, Apr. 2013.
13. Q. Li, R. Q. Hu, Y. Qian, and G. Wu, "Intracell Cooperation and Resource Allocation in a Heterogeneous Network With Relays," *IEEE Transactions on Vehicular Technology*, Vol. 62, No. 4, pp. 1770–1783, May 2013.

Chapter 2
Heterogeneous Network Model and Preliminaries

2.1 A System Model for Heterogeneous Networks

A two-tier heterogeneous network model in LTE is studied throughout this book. As shown in Fig. 2.1, consider downlink communications in a heterogeneous cellular network with high transmit power macro evolved Node Bs (MeNB) and low transmit power small evolved Node Bs (SeNB) or relay nodes (RN). Each macro cell is divided into several sectors served by directional antennas. Within each macro cell sector, there are several SeNBs or RNs uniformly distributed. Denote the total number of macro cell sectors in the system as N_c and the number of uniformly deployed SeNBs/RNs in each sector as N_r. UEs are uniformly distributed in the network with an average of N_u active UEs in each sector. The SeNBs/RNs have full radio resource management (RRM) functionalities as well as data relaying capability. Decode-and-forward relaying scheme is assumed. A UE can be either associated with a MeNB or a SeNB/RN. Denote a UE associated with a MeNB as a M-UE and a UE associated with a SeNB/RN as a S-UE/R-UE. The communication link between a MeNB and a UE is termed as a direct link, the link between a SeNB/RN and a UE as an access link, and the link between a MeNB and a SeNB/RN as a backhaul link. The backhaul link could be wired or wireless, could be ideal or non-ideal with ideal backhaul featuring a typical transmission delay of several micro seconds while non-ideal backhaul featuring a transmission delay ranging from several milli-seconds to ten of milli-seconds. For the wireless backhaul, the backhual link could be out-of-band backhaul or in-band backhaul. With in-band backhual, the backhaul link shares the same radio resource as the direct/access link.

The total frequency band can be divided into several sub-bands with each sub-band being assigned to one of the UEs. Denote the frequency-domain channel gain on the f th sub-band at time t between the i th MeNB and the k th UE as $h_{k,0,i}^{f}(t)$, and between the j th SeNB/RN in the i th sector and the k th UE as $h_{k,j,i}^{f}(t)$. The channel gain counts both long-term path loss and shadowing and short-term fading due to

R.Q. Hu and Y. Qian, *Resource Management for Heterogeneous Networks in LTE Systems*, SpringerBriefs in Electrical and Computer Engineering, DOI 10.1007/978-1-4939-0372-6__2, © The Author(s) 2014

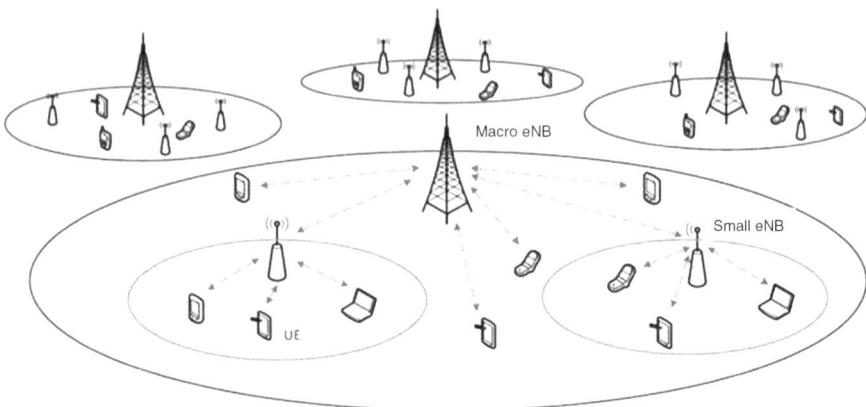

Fig. 2.1 A two-tier heterogeneous network model

multipath and mobility. The received signal-to-interference-noise-ratio (SINR) of the kth M-UE and kth S-UE/R-UE at the fth subband and tth transmission time slot can be evaluated respectively as

$$\text{SINR}_{k,0,i}^{f}(t) = \frac{P_m^f |h_{k,0,i}^f(t)|^2}{\sum_{i' \neq i} |h_{k,0,i'}^f(t)|^2 P_m^f + \sum_{i=1}^{N_c} \sum_{j=1}^{N_r} |h_{k,j,i}^f(t)|^2 P_p^f + N_0}, \quad (2.1)$$

and

$$\text{SINR}_{k,j,i}^{f}(t) = \frac{P_p^f |h_{k,j,i}^f(t)|^2}{\sum_{i=1}^{N_c} |h_{k,0,i}^f(t)|^2 P_m^f + \sum_{i=1}^{N_c} \sum_{j'=1, j' \neq j}^{N_r} |h_{k,j',i}^f(t)|^2 P_p^f + N_0}, \quad (2.2)$$

where P_m^f is the transmit power density of the MeNB at the fth subband, P_p^f is the transmit power density of the SeNB/RN at the fth subband, and N_0 is the variance of the additive noise.

The data rate in terms of bit/s/Hz for the kth UE received from the jth SeNB/RN in the ith cell on the fth radio band at time t can be calculated using Shannon formula as

$$R_{k,j,i}^{f}(t) = \log\left(1 + \text{SINR}_{k,j,i}^{f}(t)\right). \quad (2.3)$$

Data rate $R_{k,0,i}^{f}(t)$ can be similarly obtained as

$$R_{k,0,i}^{f}(t) = \log\left(1 + \text{SINR}_{k,0,i}^{f}(t)\right). \quad (2.4)$$

2.2 Mobile Associations in Heterogeneous Networks

A mobile association scheme decides the network node for a UE to connect with. In homogeneous networks, best-power based mobile association is often applied [1,2], where the kth UE is associated with the node $\mathcal{N}_{(j*,i*)}$ that it receives the highest power from, i.e.,

$$(j^*,i^*)_k = \arg \max_{i \in \{1,\cdots,N_c\}, j \in \{0,1,\cdots,N_r\}} (P_{k,j,i}|h_{k,j,i}|^2), \qquad (2.5)$$

where $P_{k,j,i}$ is the corresponding node transmission power. In heterogeneous networks, due to the transmit power disparity between a MeNB and an SeNB/RN, most of the UEs will be associated with the MeNBs if the best-power based association scheme is used. The SeNB/RN utilization will be low and the advantage of using SeNB/RN in improving the spectrum efficiency and coverage of the network could not be fully exploited. To balance the traffic load between the MeNBs and the SeNBs/RNs, range-expansion based association scheme has been proposed, which uses a bias to compensate the power difference between MeNBs and SeNBs/RNs [2], so that more UEs can be associated with SeNBs/RNs. In the range expansion based mobile association, the kth UE is associated with the best node $\mathcal{N}_{(j*,i*)}$.

$$(j^*,i^*)_k = \arg \max_{i \in \{1,\cdots,N_c\}, j \in \{0,1,\cdots,N_r\}} (|h_{k,j,i}|^2/\delta_{i,j}), \qquad (2.6)$$

where $\delta_{i,0} = 1$ and $1 < \delta_{i,j} < (P_m/P_p)$, for $j > 0$. $\delta_{i,j}$ value specifies the coverage of the macro and small cells. A small $\delta_{i,j}$ leads to a large coverage region of the small cell while a large $\delta_{i,j}$ value leads to a small coverage region of the small cell. In extreme cases, $\delta_{i,j} = 1$ corresponds to path-loss based mobile association and $\delta_{i,j} = (P_m/P_p)$ corresponds to best-power based mobile association. Figure 2.2 illustrates the different mobile association schemes.

Range-expansion based mobile association scheme can effectively expand the coverage range of small cells and therefore improves the overall system spectrum efficiency. However, with a fixed bias value, the coverage range of small cells are kept fixed. It cannot be adapted to the traffic load at small cells and therefore is less flexible and not able to fully exploit the capacity of the small cells. Mobile association schemes have been studied in heterogeneous networks for a mixed deployment of macro eNBs and small relay nodes in [3–8]. In Chap. 3, a load-balancing based mobile association scheme is presented for the two-tier heterogeneous network model in this book. The presented scheme adapts mobile association according to the macro and small cell load condition. The load-balancing based mobile association is shown effectively to improve the network spectrum efficiency as compared to best-power based and range-expansion based mobile association schemes.

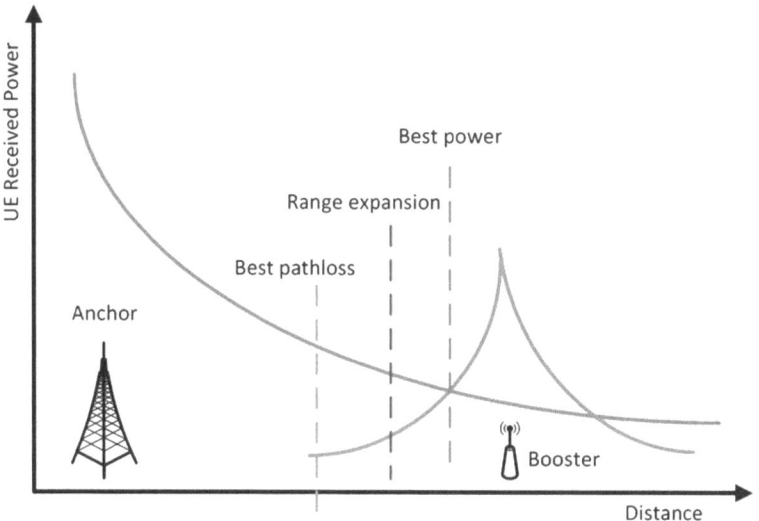

Fig. 2.2 Illustration of different mobile association schemes

2.3 Enhanced Inter-cell Interference Coordination in Heterogeneous Networks

Range expansion based or load-balancing based mobile association schemes expand the coverage range of small cells and improve the overall network spectrum efficiency. However, the UEs located at the edge of small cells would suffer from high interference from the MeNB. Proper interference management schemes can be used to mitigate the interference for cell edge UEs and improve the spectrum efficiency [9]. Inter-cell interference coordination (ICIC) is proposed in addressing the interference problem. By ICIC, proper resources coordination is conducted among interfering eNBs such that some of the eNBs give up some resource for the benefit of the other eNBs. The resource coordination can be done in time, frequency or spatial domain. Figure 2.3 demonstrates two examples of ICIC in time domain and frequency domain, where the MeNB reserves some of the subframes in time domain and resource blocks in frequency domain for use by the SeNB/RN.

In LTE Rel-8/9, ICIC is implemented among MeNBs to coordinate resource allocation in the frequency domain. Different frequency reuse options can be applied such as hard frequency reuse, fractional frequency reuse and soft fractional frequency reuse. To illustrate the different frequency reuse options, we use the example as shown in Fig. 2.4. By hard frequency reuse, the whole spectrum band is divided into subbands F_1 and F_2 with each subband being used by one of the cells. Hard frequency reuse completely eliminates the inter-cell interference, at the cost of reduced spectrum efficiency. By fractional frequency reuse, the frequency subband F_1 is used by both cells while the frequency subbands F_2 and F_3 are used by one of the cells,

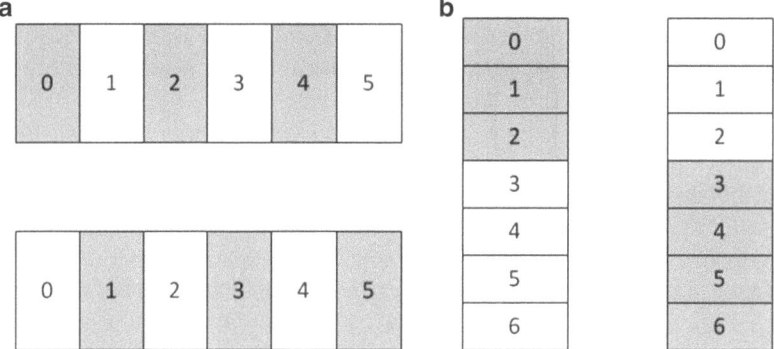

Fig. 2.3 Time domain and frequency domain frequency reuse. (**a**) Time domain resource coordination between macro and pico eNBs. (**b**) Frequency domain resource coordination between macro and pico eNBs

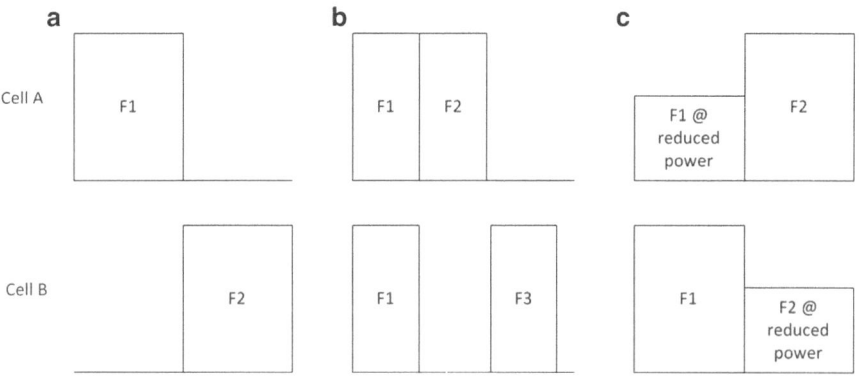

Fig. 2.4 Illustration of different frequency reuse schemes in frequency domain. (**a**) Hard frequency reuse. (**b**) Fractional frequency reuse. (**c**) Soft fractional frequency reuse

respectively. In this way, the frequency band F_1 can be used in serving the UEs at both cell centers while the subbands F_2 and F_3 can be used in serving the UEs at the edge of each cells. As the cell edge UEs are the most vulnerable to the interference, such scheme can effectively protect the cell edge UEs while improve the spectrum efficiency. By soft fractional frequency reuse, frequency bands F_1 and F_2 are used by both cells with cell A transmitting at a lower power at F_1 while cell B transmitting at a lower power at F_2. Cell A serves its inner cell UEs at F_1 and cell edge UEs at F_2 while cell B serves its inner cell UEs at F_2 and cell edge UEs at F_1. As the inner cell UEs often has good channel quality, it can still achieve a good data rate when served by a reduced DL transmission power. At the same time, the reduced eNB transmission power will cause less interference to the UEs at the edge of the neighboring cell. Soft fractional frequency reuse effectively alleviates the interference for cell edge UEs while maintains a high spectrum efficiency.

In LTE Rel-10, with the introduction of heterogeneous networks, UEs at the edge of small cells would suffer from high interference. Control channel is the most vulnerable to such macro-small interference as the control payload is typically distributed across the entire bandwidth and thus cannot be protected by the frequency domain based ICIC. As a result, UEs at the small cell edge would experience coverage holes on their control channel signals. To overcome this issue, enhanced inter-cell interference coordination (eICIC) has been proposed in LTE-Rel 10 [10]. Time domain interference coordination is applied with the introduction of almost-blank subframes (ABSF) [10]. During the ABSF, the interfering eNB does not transmit user data, but may transmit system broadcasting and reference signals, therefore the term "almost blank". The residual interference can be canceled by interference cancellation schemes at the receiver side. Resource coordination schemes have been studied in time, frequency, and power domains in a heterogeneous LTE relay network in [9] and [11]. An optimal fractional frequency reuse and power control scheme has been proposed that can effectively coordinate the interference among high power and low power nodes. In Chap. 4, an optimization framework for interference management with fractional frequency reuse is presented for the two-tier heterogeneous network model.

2.4 Intra-cell Cooperation in Heterogeneous Networks

ICIC/eICIC improves cell-edge UE performance by proper interference management. For heterogeneous networks with small coverage range expansion, a UE at the edge of a small cell would receive comparable signal quality from SeNB/RN and MeNB. It is therefore possible to enhance the cell-edge UE performance by applying intra-cell coordinated multiple point (CoMP) joint transmission from MeNB and SeNB/RN.

In LTE Rel-11, three kinds of CoMP schemes have been adopted, namely, coordinated scheduling and beamforming (CS/CB) CoMP, joint transmission (JT) CoMP, and dynamic point selection (DPS) CoMP. Figure 2.5 demonstrates the different CoMP schemes. In CS/CB, the two eNBs simultaneously transmit to their

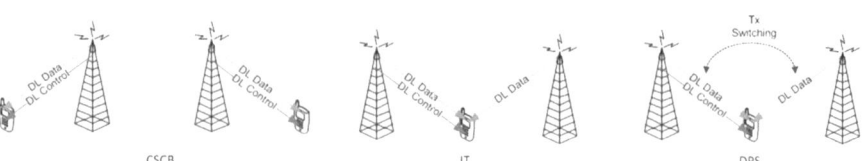

Fig. 2.5 Illustration of CoMP scheme options

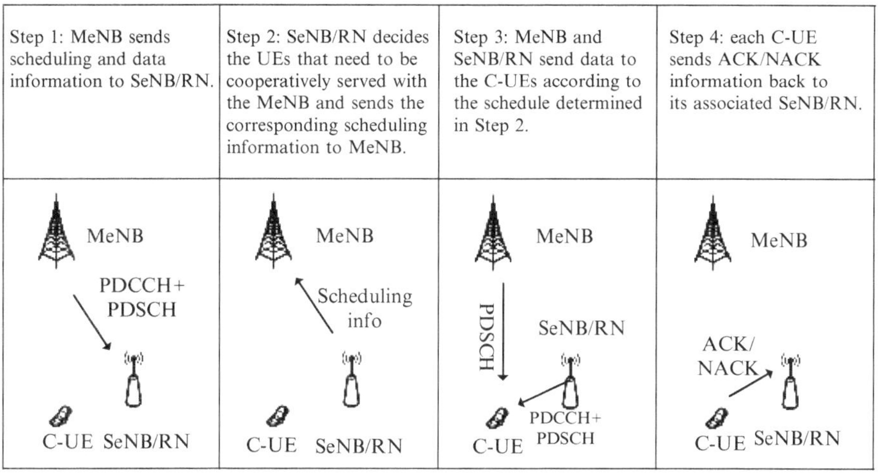

Step 1: MeNB sends scheduling and data information to SeNB/RN.	Step 2: SeNB/RN decides the UEs that need to be cooperatively served with the MeNB and sends the corresponding scheduling information to MeNB.	Step 3: MeNB and SeNB/RN send data to the C-UEs according to the schedule determined in Step 2.	Step 4: each C-UE sends ACK/NACK information back to its associated SeNB/RN.
MeNB PDCCH+ PDSCH C-UE SeNB/RN	MeNB Scheduling info C-UE SeNB/RN	MeNB PDSCH SeNB/RN PDCCH+ C-UE PDSCH	MeNB ACK/ NACK C-UE SeNB/RN

Fig. 2.6 The first intra-cell CoMP scheme

respective UEs in the same frequency resource in a coordinated way such that the mutual interference at the receiver side is minimized. In JT, each UE receives signals transmitted from both eNBs with one eNB acting as the master eNB responsible for both control and data channel transmission while the other eNB acting as slave eNB for conveying user data only. In DPS, the UE can select to receive user data from the eNB with the best channel condition, although the control signaling will remain to be received from its serving/anchored eNB.

Implementation of the above mentioned CoMP schemes assumes ideal wired backhaul between the eNBs in the CoMP set. For heterogeneous networks with wireless backhaul connection between macro and small eNBs, two possible implementations of intra-cell CoMP schemes are shown in Figs. 2.6 and 2.7. Refer to [12] for further details of the intra-cell cooperative communications in LTE systems.

In the first intra-cell CoMP scheme, there are both M-UE and S-UE/R-UE. Depending on the channel condition and available resources, MeNB can assist the communication between the SeNB/RN and some of the S-UEs/R-UEs by transmitting cooperatively with the SeNB/RN. Denote such UEs as cooperative UEs (C-UE). Figure 2.6 illustrates the intra-cell CoMP strategy for serving the C-UEs. Communications from a MeNB and an SeNB/RN to a C-UE take place in four transmission steps. In the first step, MeNB sends control and data information to SeNB/RN. Control information is sent via the physical downlink control channel (PDCCH) and the data information is sent via the physical downlink shared channel (PDSCH). In the second step, SeNB/RN decides whether to use cooperation or not for its associated S-UEs/R-UEs. If cooperation is needed, the SeNB/RN sends the corresponding scheduling and control information to the MeNB on the wireless backhaul. In the third step, upon receiving the scheduling information from the SeNB/RN, the MeNB arranges its transmission by sending data information to the

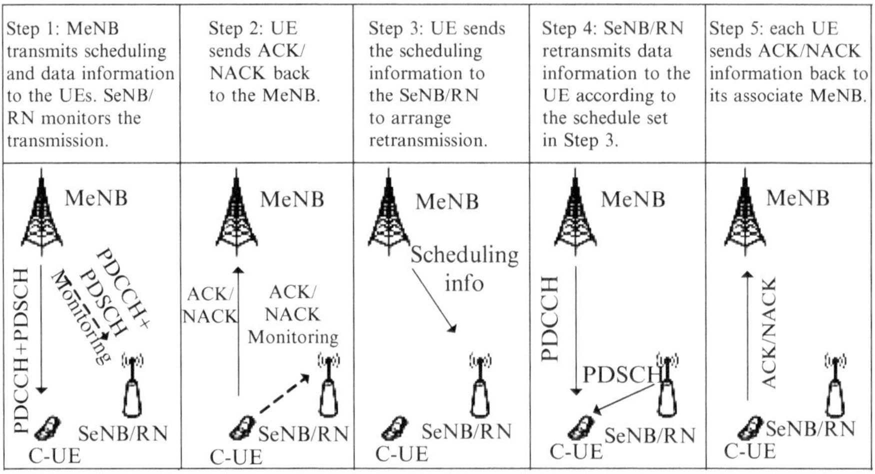

Step 1: MeNB transmits scheduling and data information to the UEs. SeNB/ RN monitors the transmission.	Step 2: UE sends ACK/ NACK back to the MeNB.	Step 3: UE sends the scheduling information to the SeNB/RN to arrange retransmission.	Step 4: SeNB/RN retransmits data information to the UE according to the schedule set in Step 3.	Step 5: each UE sends ACK/NACK information back to its associate MeNB.
MeNB PDCCH+PDSCH PDCCH+PDSCH Monitoring SeNB/RN C-UE	MeNB ACK/ NACK ACK/ NACK Monitoring SeNB/RN C-UE	MeNB Scheduling info SeNB/RN C-UE	MeNB PDCCH PDSCH SeNB/RN C-UE	MeNB ACK/NACK SeNB/RN C-UE

Fig. 2.7 The second intra-cell CoMP scheme

C-UE together with the SeNB/RN. With the received signals from the MeNB and the SeNB/RN, the C-UE decodes the information using joint decoding methods such as maximum likelihood (ML) decoding or maximum ratio combining (MRC) decoding. In the fourth step, the C-UE sends back ACK/NACK message to its associated SeNB/RN. In the first intra-cell CoMP scheme, the SeNB/RN creates a new cell with a separate cell ID distinct from the donor MeNB and appears to the UEs in the same way as a regular MeNB. Layer-3 functions are performed by the SeNB/RN.

In the second intra-cell CoMP scheme, UEs in the network are associated with the MeNB with some of them being served solely by the MeNB and the others being served with the help of the SeNB/RN. As shown in Fig. 2.7, communications take place in five transmission steps. In the first step, MeNB decides, for each M-UE, whether to serve cooperatively with the SeNB/RN or not. Based on that, MeNB sends scheduling and data information to the UEs and the SeNB/RN, respectively. The same information is sent to the C-UE and the SeNB/RN at a data rate that ensures successful decoding at the SeNB/RN. Upon receiving the data information, the UEs and the SeNB/RN decode their respective received information. In the second step, the UEs send back ACK/NACK message to the MeNB. The SeNB/RN monitors the ACK/NACK message from the C-UE. In the third step, the MeNB sends scheduling information to the SeNB/RN to arrange for retransmission from the SeNB/RN to the C-UEs that fail to decode. In the fourth step, the MeNB sends scheduling and data information to the other M-UEs. The SeNB/RN re-transmits its received data information to the C-UEs as scheduled in the third frame. Upon receiving from the SeNB/RN, the C-UE decodes its information using the received signal from the MeNB and the SeNB/RN in the first and the third steps. The rate of the re-transmitted information from the SeNB/RN is pre-determined and is set to ensure successful decoding at the C-UE. In the fifth step, the ACK/NACK message is then sent back from each C-UE to the associated MeNB. In this intra-cell CoMP scheme, the SeNB/RN is transparent to the UEs, i.e., all the scheduling instruction is sent from the MeNB, and the UEs is not aware of the existence of the SeNB/RN. The SeNB/RN does not have a cell ID and thus does not create any new cells.

In [13–15], a downlink intra-cell cooperative transmission and optimal intra-cell CoMP resource allocation schemes are explored in heterogeneous networks with cooperative relays. The schemes are optimized by selecting the best SINR threshold to form intra-cell cooperation. In Chap. 5, radio resource allocation schemes with the two-tier heterogeneous networks in LTE are presented. Radio resource allocation schemes with intra-cell CoMP and in-band wireless backhaul are studied, and an optimal framework with resource allocation strategy is presented that is asymptotically optimal on the proportional fairness metric.

References

1. K. I. Pedersen, T. E. Kolding, F. Frederiksen, D. Laselva, and P. E. Mogensen, "An overview of downlink radio resource management for UTRAN long-term evolution," *IEEE Commun. Mag.*, vol. 47, pp. 86–93, Jul. 2009.
2. S. Kyuho, C. Song, and G. Veciana, "Dynamic association for load balancing and interference avoidance in multi-cell networks," *IEEE Trans. Wireless Commun.*, vol. 8, pp. 3566–3576, Jul. 2009.
3. R. Q. Hu, Y. Yu, Z. Cai, J. E. Womack, Y. Song, "Mobile Association in a Heterogeneous Network," in *Proc. of IEEE ICC 2010*, Cape Town, South Africa, May 2010.
4. Y. Yu, R. Q. Hu, C. Bontu, Z. Cai, "Mobile Association and Load balancing in a Cooperative Relay Enabled Cellular Network," *IEEE Communications Magazines*, 49(5), pp. 83–89, May 2011.
5. Y. Yu, R. Q. Hu, Z. Cai, "Optimal Load Balancing and Its Heuristic Implementation in a Heterogeneous Relay Network," in *Proc. of IEEE GLOBECOM 2011*, Houston, Texas, Dec. 2011.
6. Q. C. Li, R. Q. Hu, G. Wu and Y. Qian, "On the optimal mobile association in heterogeneous wireless relay networks," in *Proc. of IEEE INFOCOM 2012*, Orlando, FL, Mar. 2012.
7. Q. C. Li, Y. Xu, R. Q. Hu, G. Wu, "Pricing-based distributed mobile association for heterogeneous networks with cooperative relays," in *Proc. of IEEE ICC 2012*, Ottawa, Canada, June 2012.
8. R. Chen, R. Q. Hu, "Joint uplink and downlink optimal mobile association in a wireless heterogeneous network," in *Proc. of IEEE GLOBECOM 2012*, Anaheim, CA, USA, Dec. 2012.
9. R. Q. Hu, Y. Qian, and W. Li, "On the downlink time, frequency and power coordination in an LTE relay network," in *Proc. of IEEE GLOBECOM 2011*, Houston, Texas, Dec. 2011.
10. Overview of 3GPP Release 10 V0.1.11 (2013–12)
11. Q. Li, R. Q. Hu, Y. Xu, and Y. Qian, "Optimal Fractional Frequency Reuse and Power Control in the Heterogeneous Wireless Networks," *IEEE Transactions on Wireless Communications*, vol. 12, no. 6, pp. 2658–2668, Jun. 2013.
12. Q. C. Li, R. Q. Hu, Y. Qian, and G. Wu, "Cooperative Communications for Wireless Networks: Techniques and Applications in LTE-Advanced Systems," *IEEE Wireless Commun. Mag.*, vol. 19, no. 2, pp. 22–29, Apr. 2012.

13. Y. Xu, R. Q. Hu, "Optimal Intra-cell Cooperation in the Heterogeneous Relay Network," in *Proc. of IEEE GLOBECOM 2012*, Anaheim, CA, Dec. 2012.
14. Y. Xu, R. Q. Hu, Q. C. Li, Y. Qian, "Optimal Intra-cell Cooperation with Precoding in the Wireless Heterogeneous Networks," in *Proc. of IEEE WCNC 2013*, Shanghai, China, Apr. 2013.
15. Q. Li, R. Q. Hu, Y. Qian, and G. Wu, "Intracell Cooperation and Resource Allocation in a Heterogeneous Network With Relays," *IEEE Transactions on Vehicular Technology*, Vol. 62, No. 4, pp. 1770–1783, May 2013.

Chapter 3
Mobile Association for Heterogeneous Networks

3.1 Mobile Association Scheme Based on Load-Balancing with Full Frequency Reuse

Traditional way of mobile association is based on a best-power approach where each UE associates with the eNB that it receives the highest power from. In heterogeneous networks, however, due to the disparity in transmit power of the macro and small cell eNBs, if best-power based mobile association is again used, most of the UEs will associate with the macro eNBs leaving the small cell eNBs largely under-utilized. This would undermine the ability of small cells in traffic-offloading and data rate enhancement. Mobile association schemes in heterogeneous network environment have investigated in various scenarios [1–7], as briefly reviewed in Chap. 1. In this chapter, a load-balancing based mobile association framework is presented that optimizes the mobile association by taking account of traffic load at MeNBs and RNs, the available resources of macro and small cells and the network capacity scalability. Mobile association is based on the channel state in the large scale. The received SINRs at the UEs are often used for deciding the mobile association strategy. For mobile association purpose, the received SINR is calculated from Eqs. (2.1) and (2.2) with the large-scale channel pathloss being used as the channel coefficient. Since the large scale SINR is the same for all the frequency subbands of a UE, in this chapter, the SINR is simply denoted without the frequency subband index.

The load-balancing based mobile association scheme can be formulated as an optimization problem with the objective to maximize the total number of UEs being accepted in the network while minimize the overall network resource use. A decision variable $x_{k,0,i}$ is defined to indicate the association status between the kth UE and the ith cell. Specifically,

$$x_{k,0,i} = \begin{cases} 1 & \text{if } k\text{th UE is associated with } i\text{th eNB} \\ 0 & \text{otherwise.} \end{cases} \tag{3.1}$$

R.Q. Hu and Y. Qian, *Resource Management for Heterogeneous Networks in LTE Systems*, SpringerBriefs in Electrical and Computer Engineering, DOI 10.1007/978-1-4939-0372-6_3, © The Author(s) 2014

Similarly, the decision variable $x_{k,j,i}$ is used to indicate the association status between the kth UE and the jth RN in the ith sector.

The optimization problem for achieving load-balancing mobile association is formulated as follows.

$$\max_{x_{k,j,i}} G(\mathbf{x}) = \rho_1 \sum_{i=1}^{N_c} \sum_{j=0}^{N_r} \sum_{k=1}^{N_u} x_{k,j,i} - \rho_2 \Phi \tag{3.2}$$

s.t.

$$\sum_{k=1}^{N_u} x_{k,0,i} c_{k,0,i} + \sum_{j=1}^{N_r} \sum_{k=1}^{N_u} x_{k,j,i} c_{k,j,i}^b \le C_i^M \text{ for } i = 1, \cdots, N_c \tag{3.3}$$

$$\sum_{k=1}^{N_u} x_{k,j,i}(c_{k,j,i} + c_{k,j,i}^b) \le C_j^P \text{ for } j = 1, \cdots, N_r, i = 1, \cdots, N_c \tag{3.4}$$

$$\sum_{i=1}^{N_c} \sum_{j=1}^{N_r} x_{k,j,i} = 1 \text{ or } 0 \text{ for } k = 1, \cdots, N_u \tag{3.5}$$

where $c_{k,0,i}$ denotes the radio resources needed for the kth UE being communicated with the ith MeNB, $c_{k,i,j}$ denotes the radio resources needed on the access link between the kth UE and jth RN, and $c_{k,j,i}^b$ denotes the radio resource needed on the backhaul link between the ith MeNB and the jth RN. For a SINR value $\text{SINR}_{k,0,i}$, $c_{k,0,i}$ can be calculated as

$$c_{k,0,i} = f(E\{\text{SINR}_{k,0,i}\}). \tag{3.6}$$

The function $f(\cdot)$ provides a mapping between the SINR value and the radio resource requirement. An example of the $f(\cdot)$ function can be derived from Shannon's capacity formula, which is given by

$$f(E\{\text{SINR}_{k,0,i}\}) = \frac{\Psi_k}{\log(1 + E\{\text{SINR}_{k,0,i}\})}, \tag{3.7}$$

where Ψ_k is a user specific coefficient which reflects the QoS requirement of the user. Note that as mobile association is based on the long-term pathloss and the universal frequency reuse is assumed, the frequency subband index is omitted in the SINR expression. The values of $c_{k,i,j}$ and $c_{k,j,i}^b$ can be similarly calculated. C_i^M and C_j^P denote the total available radio resources of the ith MeNB and the jth RN, respectively, and Φ is defined as the hypothetical resource consumption, defined as follows.

$$\Phi = \sum_{i=1}^{N_c} \sum_{k=1}^{N_u} x_{k,0,i} c_{k,0,i} + \sum_{i=1}^{N_c} \sum_{j=1}^{N_r} \sum_{k=1}^{N_u} x_{k,j,i} (W c_{k,j,i} + c_{k,j,i}^b), \qquad (3.8)$$

where W is a coefficient specifying the weight of the RN radio resources with respect to the MeNB radio resources in the objective function.

The weight coefficient W is crucial in the optimization of the network capacity and resource utilization. A more detailed discussion on the choice of W value will be presented later in this section. The first term in the objective function (3.2) evaluates the number of UEs being served by the network and the second term in the objective function evaluates the radio resources used in serving these UEs. The maximization of the objective function therefore requires the maximization of the total number of UEs accepted in the network and the minimization of the total radio resources consumed. This objective function properly formulates the goal in implementing mobile association in the heterogeneous network. The coefficients ρ_1 and ρ_2 specify the relative importance between the number of UEs accepted and the resource consumption. In an overloaded system, $\rho_1 > \rho_2$ stresses on capacity maximization. In an underloaded system, $\rho_1 < \rho_2$ optimizes resource utilization. Constraints (3.3) and (3.4) correspond to the resource constraints at the MeNBs and RNs, respectively. Constraint (3.5) indicates that one UE can only associate with one of the MeNBs or RNs. This constraint can be relaxed by allowing each UE connect with multiple network nodes, acquiring diversity gains. However, only one associated node per UE is considered in this chapter.

The optimization problem above is a 0-1 knapsack problem. This problem is NP-hard. An optimal solution is difficult to obtain in real time, especially given the large number of UEs in the network. A pseudo-optimal solution based on a gradient descent method is presented here. For a linear optimization problem, the pseudo-optimal solution approaches the global optimal solution which is located at the boundary of the constraint region. To apply the gradient descent method, the domain of the integer $x_{k,j,i}$ is relaxed as $x_{k,j,i} \in [0, 1]$. In this case, $x_{k,j,i}$ indicates the probability that the kth UE associates with the jth RN in the ith MeNB. Using gradient descent method, the value of $x_{k,j,i}$ is updated along the direction $\Delta x_{k,j,i} = \partial G(\mathbf{x})/\partial x_{k,j,i}$ as

$$x_{k,j,i}(t) = x_{k,j,i}(t-1) + \delta \Delta x_{k,j,i}, \qquad (3.9)$$

where δ is the step size. The value of $x_{k,j,i}$ is updated using (3.9) until the constraints in (3.3) and (3.4) are reached with equality. The $x_{k,j,i}$ values are then sorted in descending order. The $x_{k,j,i}$'s on the top of the list are those with high association probability. The UEs are accepted into the network sequentially in the order specified by the ordered $x_{k,j,i}$ list. Each UE can only be associated with one of the MeNBs or RNs. Upon the acceptance of each UE, the constraints in (3.3)–(3.5) are checked. The whole procedure stops when all the UEs are accepted

---

Algorithm 1: Pseudo-optimal solution of the mobile association problem with full frequency reuse

---

1. **Initialization**:
Set $x_{k,j,i} = 0$, acceptUEset $= \emptyset$, acceptUEnum $= 0$,
Mres $= [C_1^M, \cdots, C_{N_c}^M]$, Pres $= [C_1^P, \cdots, C_{N_r}^P]$,
$\Delta x_{k,0,i} = \partial G(\mathbf{x}) / \partial x_{k,0,i}$, $\Delta x_{k,j,i} = \partial G(\mathbf{x}) / \partial x_{k,j,i}$
2. **Update the** $x_{k,j,i}$ **values**:
For $i = 1, \cdots, N_c$ **do**
$\{$ $\Phi_i = \sum_{k=1}^{N_u} c_{k,0,i} x_{k,0,i} + \sum_{j=1}^{N_r} \sum_{k=1}^{N_u} c_{k,j,i} x_{k,j,i}$
  While $(\Phi_i < C_i^M)$ **do**
  $\{$ $x_{k,0,i} = x_{k,0,i} + \delta \Delta x_{k,0,i}$
  $x_{k,j,i} = x_{k,j,i} + \delta \Delta x_{k,j,i}$
  for $j = 1, \cdots, N_r$ **do**
  $\{$ **If** $\sum_{k=1}^{N_u} x_{k,j,i} c_{k,j,i} \geq C_j^P$
  Then $\Delta x_{k,j,i} = 0$ for $k = 1, \cdots, N_u\}$
  $\Phi_i = \sum_{k=1}^{N_u} c_{k,0,i} x_{k,0,i} + \sum_{j=1}^{N_r} \sum_{k=1}^{N_u} c_{k,j,i} x_{k,j,i}$
  $\}\}$
3. **Mobile association according to the** $x_{k,j,i}$ **values**:
$[K, J, I] = \text{Sort}(x_{k,j,i}, \text{descent})$
$k = 0$
While (acceptUEnum $<$ totalUEnum)&(Mres $> \mathbf{0}$)
$\{$ $k = k + 1$
If $(K(k) \notin$ acceptUEset$)$ **Then**
$\{$ **If** $(J(k) = 0)$&$(\text{Mres}(I(k)) - c_{K(k),0,I(k)} \geq 0)$
$\{$ acceptUEset $=$ acceptUEset $\cup \{K(k)\}$
acceptUEnum $=$ acceptUEnum $+ 1$
Mres$(I(k)) = $ Mres$(I(k)) - c_{K(k),0,I(k)}$
$\}$
If $(J(k) \neq 0)$&$(\text{Mres}(I(k)) - c_{K(k),J(k),I(k)}^b \geq 0)$&
$(\text{Pres}(J(k)) - c_{K(k),J(k),I(k)}^r - c_{K(k),J(k),I(k)}^b \geq 0)$
$\{$ acceptUEset $=$ acceptUEset $\cup \{K(k)\}$
acceptUEnum $=$ acceptUEnum $+ 1$
Mres$(I(k)) = $ Mres$(I(k)) - c_{K(k),J(k),I(k)}^b$
Pres$(J(k)) = $ Pres$(J(k)) - c_{K(k),J(k),I(k)}^r - c_{K(k),J(k),I(k)}^b \}\}\}$

---

in the network for underloaded case or all the constraints are reached for overloaded case. The detailed algorithm in finding the pseudo-optimal solution of the mobile association with full frequency reuse is shown in Algorithm 1.

Since we have

$$\frac{\partial G(\mathbf{x})}{\partial x_{k,0,i}} = \rho_1 - \rho_2 c_{k,0,i} \tag{3.10}$$

and

$$\frac{\partial G(\mathbf{x})}{\partial x_{k,j,i}} = \rho_1 - \rho_2 (W c_{k,j,i} + c_{k,j,i}^b) \tag{3.11}$$

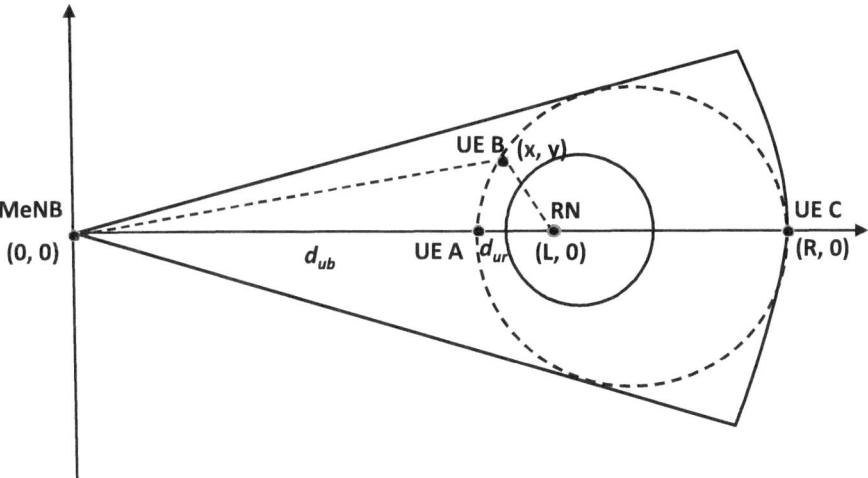

Fig. 3.1 The coverage area of the MeNB and the RN

Given the same update step, links which require low radio resources from the host node yield high association probability. Moreover, links connecting to different MeNBs have different update steps, which are determined by the number of UEs and RNs wishing to connect to the MeNB and the total available radio resources of the MeNB. A MeNB with high association requests will be saturated after a small number of iterations, and therefore the association probability with this MeNB will be low, especially for the UEs with relative high resource requirement. Therefore, high association probability happens in the two scenarios, i.e., when the required radio resource is low, or, when the number of competing UEs to a same host is low. By sequentially selecting UEs from high probability to low probability, efficient radio resource usage and balanced traffic load can be achieved.

The choice of the value of W is critical in achieving load-balancing among the MeNBs and the RNs. A low value of W reduces the weight of the RN resources in the objective function (3.2), which helps to increase the RN utility. It is therefore interesting to investigate the relationship between the W value and the portion of the UEs associated with the RN. With the assumption that the RNs are uniformly distributed along the same circle within a sector, each cellular sector is divided into M_r equal-sized sub-sectors so that only one RN resides in each sub-sector. Each sector region is approximated as a circular area with radius R. The distance between the RN and the MeNB is L. Figure 3.1 shows such an example. For a network with uniformly distributed UEs, the system-wide load balancing task can be well represented by the load balancing task within each sub-sector. Denote the number of UEs associated with the MeNB and the RN as K_{ub} and K_{ur}, respectively, and $\mathcal{K}$ as the set of UEs in the sub-sector. The relationship between K_{ub} and K_{ur} can be approximated as follows.

$$\frac{(1 + \theta_r)K_{ur}}{K_{ub} + \theta_b K_{ur}} = \beta \frac{P_r F_r E_{k \in \mathcal{K}}\{|h_{k,j,i}|^2\}}{P_b F_b E_{k \in \mathcal{K}}\{|h_{k,0,i}|^2\}} = \Gamma, \tag{3.12}$$

where θ_r indicates the average access link spectrum efficiency (SE) with respect to the average backhaul link SE and θ_b indicates the average direct link SE with respect to the average backhaul link SE, i.e., if the average direct link and the average backhaul link transmit at 1 and 10 b/s/Hz, respectively, then $\theta_b = 0.1$. Let F_r and F_b denote the frequency band the RN and the MeNB can access, respectively. In the full frequency reuse case, $F_b = F_r$. $\theta_b K_{ur}$ translates the number of UEs associated with the RN to an equivalent number of UEs associated with the MeNB, and $K_{ub} + \theta K_{ur}$ is thus the equivalent total number of UEs associated with the MeNB. Similarly, $(1 + \theta_r)K_{ur}$ is the equivalent total number of UEs associated with the RN. The parameter β indicates the extent to which the coverage range of the RN being extended by a given mobile association scheme. $\beta = 1$ corresponds to the best-power mobile association. $\Gamma = 1$ corresponds to the pass-loss based mobile association. For load-balancing based mobile association, we have

$$\beta \in \left[1, \frac{P_b F_b E\{|h_{k,0,i}|^2\}}{P_r F_r E\{|h_{k,j,i}|^2\}} \right]. \tag{3.13}$$

From (3.12), it has that

$$\frac{K_{ur}}{K_{ub}} = \frac{\Gamma}{1 + \theta_r - \Gamma \theta_b}. \tag{3.14}$$

With uniformly distributed UEs, the ratio of UE numbers can be translated into the ratio of coverage areas between the RN and the MeNB. Denote S_c as the total area of the sub-sector, S_u as the coverage area of the MeNB and S_r as the coverage area of the RN, then

$$\frac{S_r}{S_u} = \frac{S_r}{S_c - S_r} = \frac{\Gamma}{1 + \theta_r - \Gamma \theta_b}. \tag{3.15}$$

Where $S_c = \pi R^2 / 3M_r$, and S_r can be expressed as a function of W, since W is part of the association strategy and consequently impacts the coverage area of the MeNB and the RN.

To derive $S_r(W)$, note that one of the key conditions for the kth UE to associate with the jth RN in the kth sector is

$$c_{k,0,i} > W c_{k,j,i} + c^b_{k,j,i}. \tag{3.16}$$

For $c^b_{k,j,i} = \theta_b c_{k,0,i}$, UEs on the boundary between the coverage region of the MeNB and the RN satisfy

$$(1 - \theta_b)c_{k,0,i} = W c_{k,j,i}. \tag{3.17}$$

From (3.6) and (3.7), it has that

$$\frac{1 - \theta_b}{\log(1 + \text{SINR}^u_{k,0,i})} = \frac{W}{\log(1 + \text{SINR}^u_{k,j,i})}. \tag{3.18}$$

Since $\text{SINR}^u_{k,j,i}$ can be calculated as

$$\text{SINR}^u_{k,j,i} = \frac{|h_{k,j,i}|^2 P^r_j}{|h_{k,0,i}|^2 P^b_i + I_k}, \tag{3.19}$$

where I_k is the interference plus noise the kth UE received from the other MeNBs and RNs. By only considering the pathloss, $h_{k,j,i}$ can be modeled as

$$h_{k,j,i} = 1/d^\alpha_{k,j,i}, \tag{3.20}$$

where $d_{k,j,i}$ is the distance between the kth UE and the jth RN in the ith sector and α is the pathloss fading coefficient. For the kth UE located within the ith sector, $I_k \ll |h_{k,0,i}|^2 P^b_i$ is usually satisfied. The $\text{SINR}^u_{k,j,i}$ value can be approximated as

$$\text{SINR}^u_{k,j,i} = \frac{d^\alpha_{k,0,i} P^r_j}{d^\alpha_{k,j,i} P^b_i}. \tag{3.21}$$

UEs with the same $\text{SINR}^u_{k,j,i}$ value observe the same link quality and will choose the same node to associate with. The trajectory of the points with the same received SINR from the RN satisfies

$$\frac{d^\alpha_{ub} P^r_j}{d^\alpha_{ur} P^b_j} = \frac{(x^2 + y^2)^{\alpha/2} P^r_j}{((L - x)^2 + y^2)^{\alpha/2} P^b_i}, \tag{3.22}$$

where d_{ur} is the distance between UE A and the RN, d_{ub} is the distance between UE A and the MeNB, and coordinate system is shown in Fig. 3.1. The trajectory of the boundary is thus a circle expressed by the equation in the following

$$y^2 + \left(x - \frac{L}{1 - D^2}\right)^2 = \left(\frac{LD}{1 - D^2}\right)^2, \tag{3.23}$$

where

$$D = d_{ur}/d_{ub}. \tag{3.24}$$

An illustration of the boundary circle is shown by the dashed circle in Fig. 3.1. The area of the circular region associated with the RN can be calculated as

$$S_r = \pi r^2 = \frac{\pi L^2 D^2}{(1 - D^2)^2}. \tag{3.25}$$

Substituting (3.25) into (3.15), and note that

$$D = d_{ur}/(L - d_{ur}), \tag{3.26}$$

the value of d_{ur} corresponding to the desired traffic load distribution patten can be computed as

$$d_{ur} = \frac{DL}{1 + D}, \tag{3.27}$$

where

$$D = \frac{-L + \sqrt{L^2 + 4Q^2}}{2Q} \tag{3.28}$$

and

$$Q = R \sqrt{\frac{1}{3M_r} \frac{\Gamma}{1 + \theta_r + \Gamma(1 - \theta_b)}} \tag{3.29}$$

The received SINRs from the RN and from the MeNB at the Ath UE shown in Fig. 3.1 can be computed as

$$\text{SINR}^u_{A,0,i} = \frac{P^b_i/(L - d_{ur})^\alpha}{P^r_j/d^\alpha_{ur} + I_A}, \tag{3.30}$$

and

$$\text{SINR}^u_{A,j,i} = \frac{P^r_j/d^\alpha_{ur}}{P^b_i/(L - d_{ur})^\alpha + I_A}, \tag{3.31}$$

respectively. Since the Ath UE is located at the inner part of the cell,

$$I_A \leq P^b_i/(L - d_{ur})^\alpha \tag{3.32}$$

with high probability, setting $I_A = 0$ in (3.31) gives a good approximation for the value of $\text{SINR}^u_{A,j,i}$. For the value of $\text{SINR}^u_{A,0,i}$, since the relative value between P^r_j/d^α_{ur} and I_A is unclear, setting $I_A = 0$ only gives an upper bound for $\text{SINR}^u_{A,0,i}$. Substituting (3.30) and (3.31) with $I_A = 0$ into (3.18), the weight coefficient can be found as

$$W = \frac{(1 - \theta_b)^+ \log(1 + \text{SINR}^u_{A,j,i})}{\log(1 + \text{SINR}^u_{A,0,i})}, \tag{3.33}$$

where

$$(x)^+ = \max\{0, x\}. \tag{3.34}$$

Equations (3.27)–(3.33) give the relationship between W and Γ. It will be shown later that W is a decreasing function of Γ. Recall that Γ evaluates the ratio between the number of UEs associated with the RN and the number of UEs associated with the MeNB. As Γ increases, more UEs are associated with the RN, and the W value should be small to achieve this.

3.2 Mobile Association Scheme Based on Load Balancing with Partial Frequency Reuse

In the full frequency reuse case, the high transmit power of the MeNBs renders a low coverage area of the RNs. The coverage region of the RN can be extended by applying range extension or load-balancing based mobile association as presented in the last session. However, the UEs located at the extended boundary of the RNs may suffer high interference from the neighboring MeNBs. To overcome this inter-cell interference and make more efficient usage of the RN resources, partial frequency reuse can be applied to assign RNs and MeNBs with orthogonal frequency subbands [3]. Due to the in-band backhaul assumption, the frequency reuse scheme needs to be tailored for accommodating the backhual transmission and the direct/access link transmission. In the following, a mobile association scheme with partial frequency reuse is presented.

The RNs operate in a half-duplex TDD mode to avoid self-interference, i.e., the backhaul and access links are time division multiplexed, as shown in Fig. 3.2. In T_1, MeNB transmits over the whole frequency band with F_{11} being used to transmit to its associated UEs in the direct link and F_{12} being used to transmit to the RNs via the backhaul links. In T_2, the total frequency band is divided into two sub-bands, namely F_{21} and F_{22}. F_{21} is used by the MeNB to communicate with its associated UEs in the direct link. F_{22} is used by the RNs to convey their received information from the MeNB in T_1 to its associated UEs in the RN access link.

With partial frequency reuse, UEs associated with RNs receive no interference from the MeNBs. Following the same framework as in the full frequency reuse case, the mobile association problem can be formulated as follows.

$$\max_{x_{k,j,i}} \rho_1 \sum_{i=1}^{N_c} \sum_{j=1}^{N_r} \sum_{k=1}^{N_u} x_{k,j,i} - \rho_2 \Phi \qquad (3.35)$$

s.t.

$$\sum_{k=1}^{N_u} x_{k,0,i} c_{k,0,i} + \sum_{j=1}^{N_r} \sum_{k=1}^{N_u} x_{k,j,i} c_{k,j,i}^b \leq C_1 + C_{21} \quad \text{for } i = 1, \cdots, N_c$$

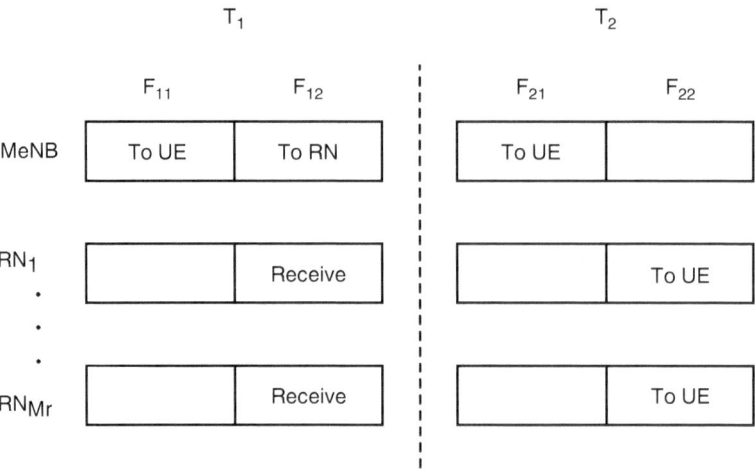

Fig. 3.2 The frequency reuse scheme

$$\sum_{k=1}^{N_u} x_{k,j,i} c_{k,j,i} \leq C_{22} \text{ for } j = 1, \cdots, N_r \tag{3.36}$$

$$\sum_{i=1}^{N_c} \sum_{j=1}^{N_r} x_{k,j,i} = 1 \text{ or } 0 \text{ for } k = 1, \cdots, N_u, \tag{3.37}$$

where,

$$C_1 = \frac{T_1}{T_1 + T_2} C, \tag{3.38}$$

$$C_2 = \frac{T_2}{T_1 + T_2} C, \tag{3.39}$$

$$C_{21} = \frac{F_{21}}{F_{21} + F_{22}} C_2, \tag{3.40}$$

$$C_{22} = \frac{F_{22}}{F_{21} + F_{22}} C_2, \tag{3.41}$$

and C denotes the total available radio resources assigned to each sector.

The gradient descent method can be used again to find the pseudo-optimal solution for the problem formulated above. The procedure is similar to the one used in the full frequency reuse case. The domain of $x_{k,j,i}$ can be relaxed to be $x_{k,j,i} \in [0, 1]$ and be viewed as the association probability between the kth UE and the jth RN in the ith MeNB. Gradient descent method is used to update the association probability until the resource constraints are met. The association links are then established by sequentially accepting the UEs according to their probabilities in a descending order. Unlike in the full frequency reuse case, where both RNs and MeNBs have access to the full radio resources, in the partial frequency reuse case, the mobile association and frequency partition between the MeNB and RN will be jointly optimized in the proposed scheme. The algorithm can be simply modified to handle the fixed frequency partition as well. The detailed algorithm in finding the pseudo-optimal mobile association scheme with partial frequency reuse is given in Algorithm 2. In the algorithm, T_1 and T_2 are fixed, and flexible partition between F_{11} and F_{12} as well as between F_{21} and F_{22} are allowed to optimize the system capacity. However, All sectors will adopt the same frequency partition to minimize inter-cell interference.

To find a good partition between F_{21} and F_{22}, start with a pre-assigned amount of resources. When the MeNB becomes overloaded first, part of the RN resources will be re-assigned to the MeNB. When the RN becomes overloaded first, MeNB will transfer part of its resources to the RN.

3.3 Online Mobile Association Algorithm

The mobile association schemes discussed in the last two sections are based on an offline approach with joint optimization among the UEs in the network. In this section, a generic online algorithm is presented by applying the gradient descent method to a batch of incoming UEs. The default batch size can be set to 1. In the following, the full frequency reuse scenario is used in describing the online algorithm. A similar online algorithm can be deduced for the partial frequency reuse case.

An online mobile association scheme is implemented for a batch of incoming UEs over a reduced scope of entire system resources, as shown in Algorithm 3. The scope is reduced since there are UEs already served by the network and already taking resources away. So the association decision for the incoming UEs can only be done based on the remaining network resources.

For the special case where the input UE set is the whole pool of UEs, then the Algorithm 3 is the same as the Algorithm 1. In another special case where the input UE set has only one UE, Algorithm 3 finds the association scheme for each of the UEs coming into the network. In general, the MeNBs and RNs can wait until a batch of UEs come into the network and do mobile association jointly for them. As demonstrated from the performance results below, the system capacity also increases as the size of UE batch increases. This improvement in network capacity comes at the cost of higher waiting time for the UEs. There is a tradeoff between the capacity and the waiting time. In reality, the right batch size can be chosen for the on-line algorithm according to the network status and service requirements. For example, when the association occurs for the UEs from idle to active transition, a relatively longer waiting time can be tolerant, so a larger batch size can be chosen to improve the network capacity. During handover for delay-sensitive applications, however, UE set size 1 should be chosen to minimize the waiting time and to guarantee QoS.

Algorithm 2: Pseudo-optimal solution of the mobile association problem with partial frequency reuse

---

1. **Initialization**:
   Set $x_{k,j,i} = 0$, acceptUEset $= \emptyset$, acceptUEnum $= 0$,
   Mallocres $= C_2 \mathbf{1}_{N_c \times 1}$, Pallocres $= C_2 \mathbf{1}_{N_r \times 1}$,
   Mres $= C_1 \mathbf{1}_{N_c \times 1} +$ Mallocres, Pres $=$ Pallocres,
   $\Delta x_{k,0,i} = \partial G(\mathbf{x})/\partial x_{k,0,i}$, $\Delta x_{k,j,i} = \partial G(\mathbf{x})/\partial x_{k,j,i}$

2. **Update the $x_{k,j,i}$ values and find the optimal radio resource allocation**:
   For $i = 1, \cdots, N_c$ **do**
   $\{ \ \Phi_i = \sum_{k=1}^{N_u} c_{k,0,i} x_{k,0,i} + \sum_{j=1}^{N_r} \sum_{k=1}^{N_u} c_{k,j,i} x_{k,j,i}$
   $\Phi_i^b = \sum_{j=1}^{N_r} \sum_{k=1}^{N_u} c_{k,j,i} x_{k,j,i}$
   $D_i = \text{Mres}(i) - \Phi_i$
   $D_i^b = C_1 - \Phi_i^b$
   P_full $= 0$
   While $(D_i > 0)$ & $(D_i^b > 0)$ **do**
   $\{ \ x_{k,0,i} = x_{i,0,i} + \delta \Delta x_{k,0,i}$
   $x_{k,j,i} = x_{i,j,i} + \delta \Delta x_{k,j,i}$
   $\Phi_i = \sum_{k=1}^{N_u} c_{k,0,i} x_{k,0,i} + \sum_{j=1}^{N_r} \sum_{k=1}^{N_u} c_{k,j,i} x_{k,j,i}$
   $\Phi_i^b = \sum_{j=1}^{N_r} \sum_{k=1}^{N_u} c_{k,j,i} x_{k,j,i}$
   $D_i = \text{Mres}(i) - \Phi_i$
   $D_i^b = C_1 - \Phi_i^b$
   $L = \max_j \sum_{k=1}^{N_u} x_{k,j,i} c_{k,j,i} - \text{Pres}(j)$
   If $(L > 0)$&$(D_i > 0)$
   $\{$ Pallocres $=$ Pallocres $+ \min\{D_i, L\} \mathbf{1}_{N_r \times 1}$
   Mallocres $=$ Mallocres $- \min\{D_i, L\} \mathbf{1}_{N_c \times 1}$
   Mres $= C_1 \mathbf{1}_{N_r \times 1} +$ Pallocres
   Pres $=$ Mallocres
   P_full $= 1\}$
   If $(D_i < 0)$ & (P_full $= 0$)
   $\{$ Mallocres $=$ Mallocres $+ \min\{|D_i|, |L|\} \mathbf{1}_{N_c \times 1}$
   Pallocres $=$ Pallocres $- \min\{|D_i|, |L|\} \mathbf{1}_{N_c \times 1}$
   Mres $= C_1 \mathbf{1}_{N_c \times 1} +$ Mallocres
   Pres $=$ Pallocres
   $D_i = \text{Mres}(i) - \Phi_i \}\}\}$
   allocRes $=$ Majority(Mallocres)
   Mallocres $=$ allocRes$\mathbf{1}_{N_c \times 1}$
   Pallocres $= (C_2 - \text{allocRes}) \mathbf{1}_{N_r \times 1}$
   Mres $= C_1 \mathbf{1}_{N_c \times 1} +$ Mallocres
   Pres $=$ Pallocres

3. **Mobile association according to the $x_{k,j,i}$ values**
   Mobile association follows the same procedure as in the full frequency reuse case.

3.4 Performance Results and Discussions

In this section, numerical results are presented to demonstrate the performance of the load-balancing mobile association scheme. The evaluation is done for a cellular network with a 19-cell 3-sector three-ring hexagonal cell structure. Two RNs are

---

Algorithm 3: Online mobile association algorithm based on gradient descent method

---

1. **Initialization**:
acceptUEset $= \emptyset$, acceptUEnum $= 0$, $M_u = |\mathscr{S}_u|$, $k = 0$,
BwPortion $= \max\{100, \text{MaxacceptUEnum}/M_u\}$,
MresP $= [C_1^M, \cdots, C_{N_c}^M]/\text{BwPortion}$,
PresP $= [C_1^P, \cdots, C_{N_r}^P]/\text{BwPortion}$,
Mres $=$ MresP, Pres $=$ PresP,
2. **Online mobile association**:
For each incoming set of UEs $\mathscr{S}_u$ **do**
$\{\ k = k + M_u$
  (UEset, UEnum, Mres, Pres)
    $=$ MAgradient$(\mathscr{S}_u, \text{Mres}, \text{Pres})$
  acceptUEset $=$ acceptUEset $\cup$ UEset
  acceptUEnum $=$ acceptUEnum $+$ UEnum
  if $(\text{mod}(k, \max\{100, M_u\}) = 0)$
  $\{$ Mres $=$ Mres $+$ MresP
    Pres $=$ Pres $+$ PresP$\}\}$

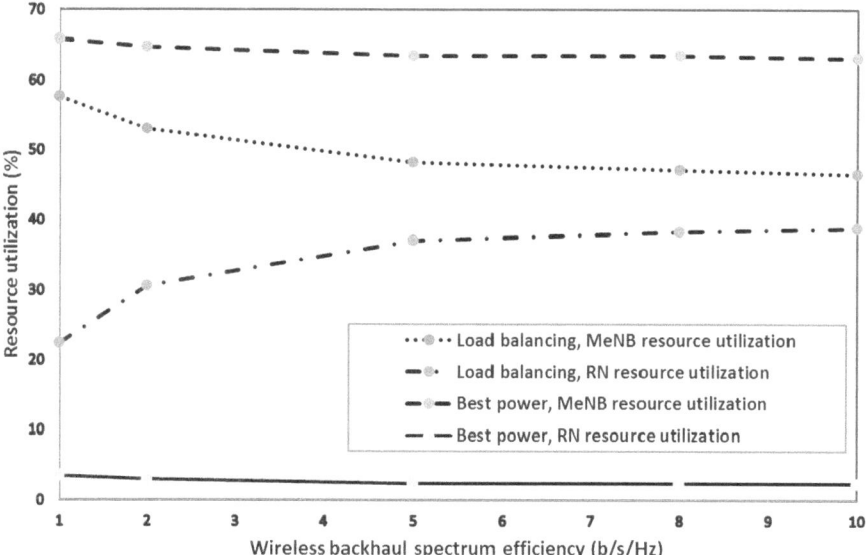

Fig. 3.3 Resource consumption comparison with full frequency reuse

uniformly deployed in each sector. Simulation setup follows the guidelines for Case 1 described in the 3GPP technical reports [8]. Transmit power of the MeNB is 46 dBm (40 W) and transmit power of the RN is 30 dBm (1 W). The UEs are uniformly distributed in the network. Each UE represents an adaptive multi-rate based VoIP user with a rate of 8.6 kbps.

In Fig. 3.3, the resource consumption at the MeNBs and the RNs are compared for the proposed and best-power based mobile association schemes. An under-

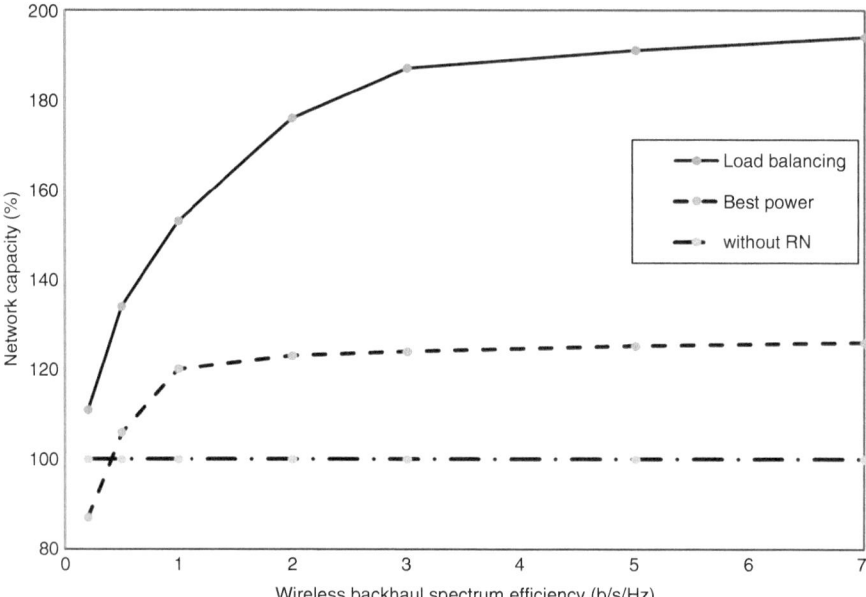

Fig. 3.4 Network capacity comparison with full frequency reuse

loaded scenario is simulated with an average of 70 UEs per sector. The resource consumption between different mobile association schemes when serving the same number of UEs in the network can be compared. A better association scheme results in lower or more balanced resource consumption and thus has higher potential in accepting more incoming UEs. As shown in Fig. 3.3, with best-power association, the resource consumption at the RNs is very low. Most of the UEs are associated with the MeNBs due to MeNBs' higher transmit power and larger coverage area. Since all the UEs in the network require resources from the MeNBs either via the direct link or the backhaul link, the high consumption of the MeNB resources leaves little room to accommodate future UEs. For the proposed scheme, balanced resource consumption at the MeNBs and at the RNs is observed. As the RNs share a larger portion of the traffic burden, less resources are used at the MeNBs. Consequently, the system is capable to accept more UEs in the future.

In Fig. 3.4, the network capacity or the maximum number of UEs accepted per sector is simulated for different association schemes under an overloaded condition. All the capacity numbers are expressed as the relative percentage of the capacity number in a network without RN. It can be seen that the proposed load-balancing based association scheme yields the highest network capacity. The network capacity increases as the backhaul link quality improves. As the backhaul link quality improves, more UEs can be associated with the RNs and more resources at the MeNBs can be released for supporting the backhaul transmission, which in turn allows the network to accept more UEs. In a HetNet with in-band backhaul, the

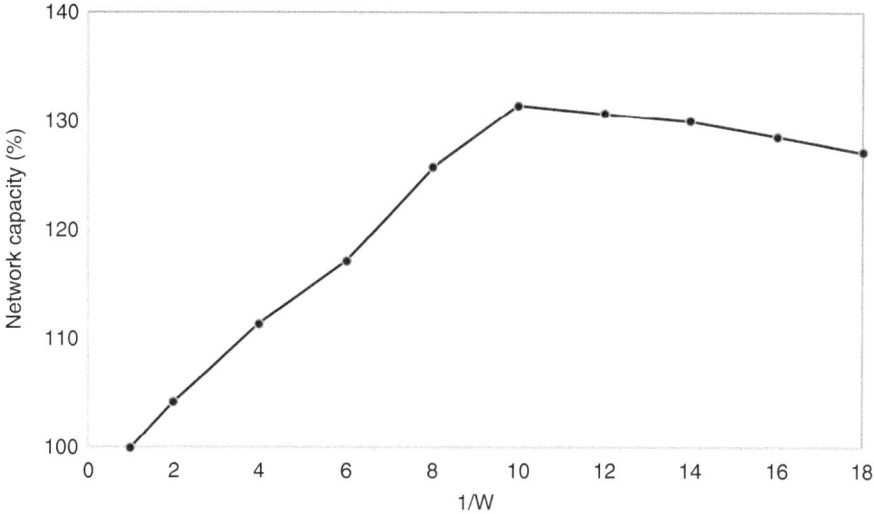

Fig. 3.5 Network capacity versus weight coefficient values, 200 UEs per sector, $\theta_b = 0.15$, $\theta_r = 0.012$

MeNB is usually the bottleneck that restricts the sector capacity and thus the most efficient use of the MeNB resources should be ensured. In contrast, the network capacity obtained with the best-power association scheme does not improve that much with respect to the backhaul link quality improvement. When the backhaul can support a data rate of 7 bits/s/Hz, a 90 % more network capacity can be achieved by the proposed scheme as compared to the best-power based scheme. Moreover, when the backhaul link quality is poor, the capacity obtained by the best-power based association could be even lower than that in the no RN case. The reason is that with best-power association, some of the UEs will be associated with the RNs regardless of backhaul condition. Due to the poor backhual link quality, relaying UE information will take more resources than direct transmission, which lowers the overall spectrum efficiency. Usually, range-expansion based association scheme may lead to low SINR for some of the UEs at the extended cell boundary under full frequency reuse. The proposed algorithm converges in less than ten iterations to the optimal solution, which is quite efficient.

The weight coefficient W is critical for the overall performance. Figure 3.5 shows that there is an optimal value W^* that maximizes the network capacity. When $W < W^*$, the cost of using RN is too high, which leads to an insufficient usage of the RN and low overall spectrum efficiency. When $W > W^*$, the cost of using RN is too low, which allows more UEs to be associated with the RN, including these UEs that have very poor SINR values. In both cases, the overall spectrum efficiency goes down. Figure 3.6 shows the relationship between W and the parameter Γ. The points of the optimal (W, Γ) pairs are marked in asterisk. Recall that Γ indicates the percentage of UEs associated with the RN, which can be

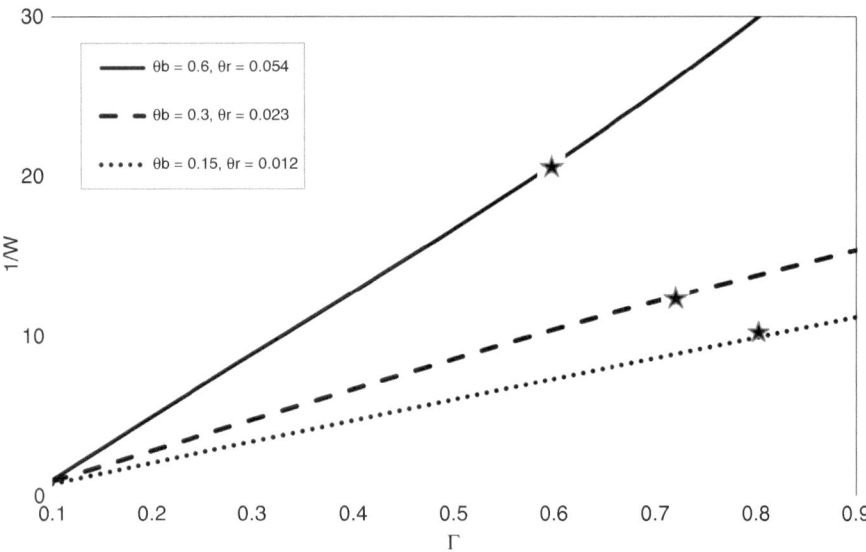

Fig. 3.6 Relationship between W and Γ

calculated as $\beta \frac{P_r F_r E\{|h_{k,j,i}|^2\}}{P_b F_b E\{|h_{k,0,i}|^2\}}$. It can be seen that W is a decreasing function of Γ. This is coincident with the understanding that in order for more UEs to be associated with the RN, W should be small. Moreover, as θ_b and θ_r decreases, i.e., as backhaul link improves, the optimal Γ increases, indicating that the optimal network capacity is achieved when more UEs associated with the RNs.

Figure 3.7 compares the network capacity of the load-balancing based, the best-power based and the pass-loss based mobile association schemes with partial frequency reuse. Path-loss based scheme is added into comparison for this case since this scheme can only work well under the partial frequency reuse. The network capacity of the load-balancing based scheme under full frequency reuse is also presented for comparison. It can be observed that as the backhaul link quality improves, network capacity is improved by applying partial frequency reuse. This is because partial frequency reuse eliminates mutual interference between MeNB and RN and improves the access link quality. Among all schemes, the proposed scheme with partial frequency reuse consistently demonstrates the highest capacity when backhaul link is better than 2 b/s/Hz. With a good backhaul link, the composite link between the UE and the MeNB via the RN may consume less resources than the direct link. The network capacity can be improved by associating more UEs with the RN. This understanding can be further verified by the curve depicting the portion of resources F_{21}. It can be seen that as the backhaul link improves, F_{21} decreases and the RNs are allocated with more resources, indicating that more UEs are associated with the RNs. Partial frequency reuse improves significantly the

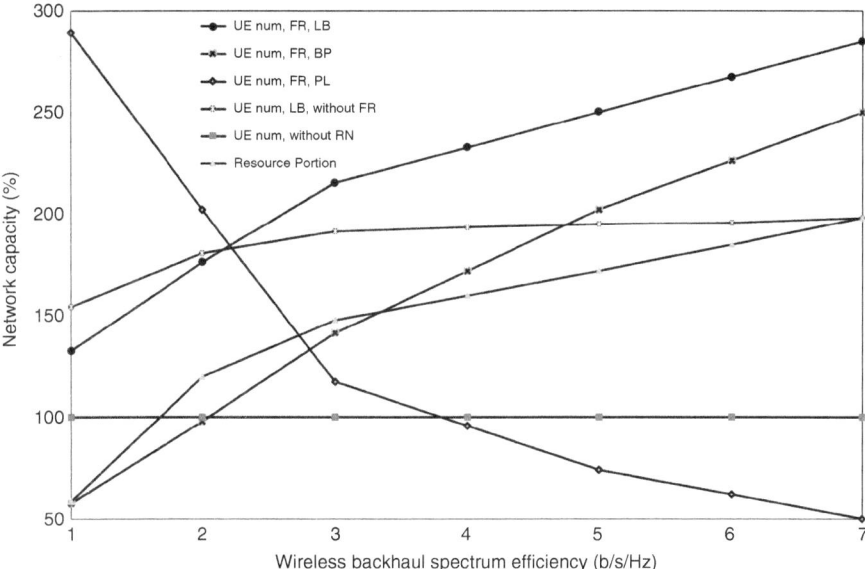

Fig. 3.7 Network capacity comparison with partial frequency reuse

network capacity of the pass-loss based algorithm compared with full frequency reuse. Without partial frequency reuse, the UEs in the extended range of the RNs will suffer very high interference from the MeNBs and the capacity of the pass-loss based algorithm could be even worse than the no RN case. However, the pass-loss based scheme with partial frequency reuse still gives a lower network capacity than the other two schemes. There are only 2 RNs per sector and path-loss based scheme extends RN coverage area beyond its power reaching limit. Adding more RNs can help to improve this situation. With backhaul data rate at 5 b/s/Hz, the median SINR of the proposed load-balancing based algorithm and the best-power based algorithm are 10.5 and 11.3 dB, respectively, while the proposed scheme achieves 32 % higher capacity than the best-power based scheme.

Figure 3.8 shows the performance of the proposed online mobile association algorithm with different sizes of the jointly processed set of UEs. The simulation is performed with a backhaul link data rate at 7 b/s/Hz. It can be seen that as the set size increases, the network capacity also increases. In an extreme case where all the UEs are jointly processed, the network capacity achieves the maximum value and corresponds to the pseudo-optimal solution obtained by Algorithm 1. Another extreme is the real-time online association where mobile association is implemented for each incoming UE. By comparing the network capacity obtained under the two extremes, it can be seen that the single-UE based online processing only lost 13 % capacity compared with the offline scheme. On the other hand, it gives much smaller processing delay and much lower waiting delay.

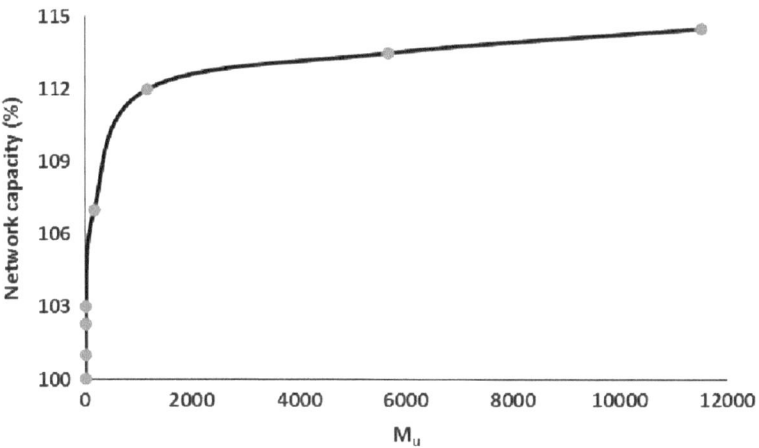

Fig. 3.8 Network capacity of the online algorithm

3.5 Summary

This chapter presented a load-balancing based mobile association framework for heterogeneous networks with wireless in-band backhaul. It has shown that balancing the traffic load between the MeNBs and the RNs is essential in improving the heterogeneous network capacity. The load-balancing based mobile association framework is proposed in the heterogeneous networks under different frequency reuse schemes. The pseudo-optimal solutions for the proposed mobile association frameworks are derived based on the gradient descent method. The advantage of the proposed mobile association in improving the network capacity has been verified by numerical results. An online algorithm which allows real-time implementation of the proposed mobile association is also developed. Performance results show that the online algorithm achieves a good tradeoff between capacity and association delay.

References

1. R. Q. Hu, Y. Yu, Z. Cai, J. E. Womack, Y. Song, "Mobile Association in a Heterogeneous Network," in *Proc. of IEEE ICC 2010*, Cape Town, South Africa, May 2010.
2. Y. Yu, R. Q. Hu, C. Bontu, Z. Cai, "Mobile Association and Load balancing in a Cooperative Relay Enabled Cellular Network," *IEEE Communications Magazines*, 49(5), pp. 83–89, May 2011.
3. R. Q. Hu, Y. Qian, and W. Li, "On the downlink time, frequency and power coordination in an LTE relay network," in *Proc. of IEEE GLOBECOM 2011*, Houston, Texas, Dec. 2011.
4. Y. Yu, R. Q. Hu, Z. Cai, "Optimal Load Balancing and Its Heuristic Implementation in a Heterogeneous Relay Network," in *Proc. of IEEE GLOBECOM 2011*, Houston, Texas, Dec. 2011.

5. Q. Li, R. Q. Hu, G. Wu and Y. Qian, "On the optimal mobile association in heterogeneous wireless relay networks," in *Proc. of IEEE INFOCOM 2012*, Orlando, FL, Mar. 2012.

6. Q. C. Li, Y. Xu, R. Q. Hu, G. Wu, "Pricing-based distributed mobile association for heterogeneous networks with cooperative relays," in *Proc. of IEEE ICC 2012*, Ottawa, Canada, Jun. 2012.

7. R. Chen, R. Q. Hu, "Joint uplink and downlink optimal mobile association in a wireless heterogeneous network," in *Proc. of IEEE Globecom 2012*, Anaheim, CA, USA, Dec. 2012

8. 3GPP TR36.814, "Further advancements for E-UTRA physical layer aspects," v9.0.0, Mar. 2010.

Chapter 4
Interference Management in Heterogeneous Networks with Fractional Frequency Reuse

4.1 Problem Formulation

As illustrated in Fig. 4.1, in an FFR scheme, the total sub-bands F is divided into two parts, f_1 and f_2 with size F_1 and F_2, respectively. In f_1, MeNBs transmit at a reduced power αP_m ($0 < \alpha < 1$) to the cell center UEs while SeNBs transmit at the full power to the UEs located at the edge of the small cells. In f_2, both MeNBs and SeNBs transmit at their respective full powers. MeNBs transmit to the cell edge UEs while SeNBs transmit to the small cell center UEs. A frequency partition coefficient β is defined as $\beta = F_1/F$.

The optimal FFR scheme is designed with the target to maximize the long-term system throughput as well as to ensure good user experience. Towards this end, it needs to decide (1) the optimal partition of the frequency sub-bands f_1 and f_2, i.e., the value of β; (2) the optimal transmit power of the MeNBs in the f_1 sub-band, i.e., the value of α. The optimization of the FFR scheme is closely related to the mobile association scheme used. A joint optimization on the FFR scheme and the mobile association scheme is preferable from performance perspective. However, such a joint optimization would be highly complicated, mathematically intractable and impractical for implementation. Resource coordination schemes have been studied in time, frequency, and power domains in a heterogeneous LTE relay network [1, 2]. An optimal fractional frequency reuse and power control scheme has been proposed that can effectively coordinate the interference among high power and low power nodes. This chapter presents an optimal FFR design under a given mobile association scheme. The mobile association scheme is decided offline, based on which a jointly optimization on (1) and (2) is conducted.

A decision variable $x_{k,0,i}^{f_1}$ is used to indicate the association status between the kth UE and the ith MeNB on its f_1 sub-bands. Specifically,

$$x_{k,0,i}^{f_1} = \begin{cases} 1 & \text{if } k\text{th UE is associated with MeNB } i \text{ on } f_1 \\ 0 & \text{otherwise.} \end{cases} \quad (4.1)$$

R.Q. Hu and Y. Qian, *Resource Management for Heterogeneous Networks in LTE Systems*, SpringerBriefs in Electrical and Computer Engineering, DOI 10.1007/978-1-4939-0372-6_4, © The Author(s) 2014

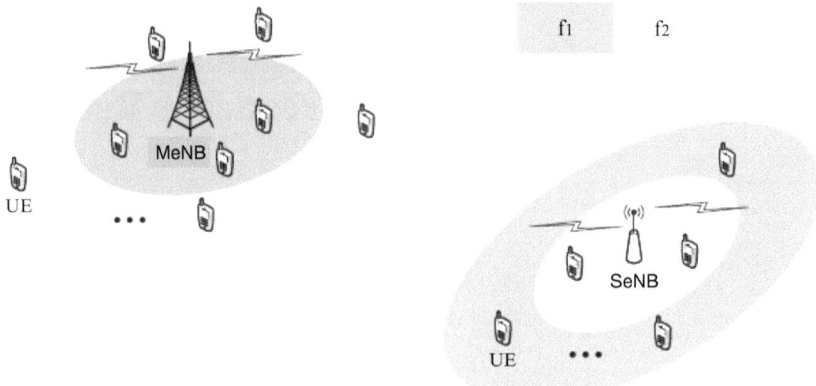

Fig. 4.1 A fractional frequency reuse scheme

The decision variables $x_{k,0,i}^{f_2}$, $x_{k,j,i}^{f_1}$ and $x_{k,j,i}^{f_2}$ are similarly defined. Note that each UE can only associate with either one MeNB or one SeNB in f_1 or f_2 subband, i.e.,

$$\sum_{i=1}^{N_c}\sum_{j=0}^{N_r}\sum_{m=1}^{2} x_{k,j,i}^{f_m} = 1. \tag{4.2}$$

Range-expansion or load-balancing based mobile association can be used in deciding the MeNB and SeNB to which the UE is associated with. A further decision on the associated area, i.e., cell center or cell edge, should be made by considering the subband partition β. Specifically, the UEs associated with each MeNB or SeNB are sorted in a descending order according to their SINRs. For the UEs associated with the MeNB, the first β portion of the sorted UEs are assigned with the subband f_1 and the remaining $(1-\beta)$ portion of the sorted UEs are assigned with the subband f_2. For the UEs associated with the SeNB, the first $(1-\beta)$ portion of the sorted UEs are assigned with the subband f_1 and the remaining β portion of the sorted UEs are assigned with subband f_2.

Based on the mobile association strategy, the optimal FFR scheme can be designed. For the optimal FFR scheme, minimizing the interference and maximizing the spectrum efficiency have normally been considered as the primary objectives. Fairness among the UEs can be another important design objective if user experience needs to be considered. Several schemes have been proposed to address the fairness issue, including the max-min fairness scheme proposed in [3], the proportional fairness scheme proposed in [4] and the competitive fairness scheme proposed in [5]. In this chapter, proportional fairness is applied by defining the sum of log-scale throughput as the performance metric.

The optimization of FFR scheme is pseudo-static, i.e., the decision is based on the long-term statistics instead of short-term information. This is consistent with the most FFR scheme designs in wireless networks. Thus the resources allocated to each UE defined in the following are the average allocation over a certain time period. So is the throughput. For UE k associated with ith MeNB cell center (or cell edge), the allocated resources in the unit of sub-band are denoted as $n_{k,0,i}^{f_1}$ (or $n_{k,0,i}^{f_2}$). Similarly, for UE k associated with the cell center (or cell edge) of the jth SeNB in the coverage of the ith MeNB, the allocated resources in the unit of subband are denoted as $n_{k,j,i}^{f_1}$ (or $n_{k,j,i}^{f_2}$). The optimization problem can be formulated as follows.

$$[\mathbf{P}_1] \quad \min_{\alpha,\beta,n_{k,j,i}^{f_m}} U(\alpha,\mathbf{n}) = -\sum_{m=1}^{2}\sum_{i=1}^{N_c}\sum_{j=0}^{N_r}\sum_{k=1}^{N_u} x_{k,j,i}^{f_m} \log(n_{k,j,i}^{f_m} R_{k,j,i}^{f_m}) \qquad (4.3)$$

subject to

$$\sum_{k=1}^{N_u} x_{k,j,i}^{f_1} n_{k,j,i}^{f_1} \leq \beta F, \text{ for } i = 1,\cdots,N_c, j = 0,\cdots,N_r \qquad (4.4)$$

$$\sum_{k=1}^{N_u} x_{k,j,i}^{f_2} n_{k,j,i}^{f_2} \leq (1-\beta)F, \text{ for } i = 1,\cdots,N_c, j = 0,\cdots,N_r \qquad (4.5)$$

$$\sum_{m=1}^{2}\sum_{i=1}^{N_c}\sum_{j=0}^{N_r} x_{k,j,i}^{f_m} n_{k,j,i}^{f_m} R_{k,j,i}^{f_m} \geq R_{\min}, \text{ for } k = 1,\cdots,N_u \qquad (4.6)$$

$$\sum_{i=1}^{N_c}\sum_{j=0}^{N_r} (x_{k,j,i}^{f_1} + x_{k,j,i}^{f_2}) \leq 1, \text{ for } k = 1,\cdots,N_u \qquad (4.7)$$

$$n_{k,j,i}^{f_t} \geq 0, \text{ for } t = 1,2, i = 1,\cdots,N_c, j = 1,\cdots,N_r \qquad (4.8)$$

$$0 \leq \alpha \leq 1 \qquad (4.9)$$

$$0 \leq \beta \leq 1. \qquad (4.10)$$

The value of $n_{k,j,i}^{f_m} R_{k,j,i}^{f_m}$ gives the average throughput for the kth UE by being associated with $\mathcal{N}_{j,i}$ in the f_m band of the ith sector and being allocated with $n_{k,j,i}^{f_m}$ subbands. $R_{k,j,i}^{f_m}$ is defined in (2.3). $n_{k,j,i}^{f_m}$ can be a non-integer, which represents the time-averaged number of subbands allocated for UE k. In the objective function (4.3), instead of directly maximizing the total throughput, the total log-scaled throughput can be maximized to achieve the proportional fairness. Note that

$$\nabla_{\mathbf{n}} U(\alpha,\mathbf{n}) = \left[\frac{\partial U(\alpha,\mathbf{n})}{n_{k,j,i}^{f_m}}\right]_{k,j,i}^{f_m} = -\frac{x_{k,j,i}^{f_m}}{n_{k,j,i}^{f_m}}. \qquad (4.11)$$

For a large $n_{k,j,i}^{f_m}$, a further increase in its value will lead to only a marginal increase in the total log-scaled throughput. To maximize the total log-scaled throughput, it is more beneficial to increase these $n_{k,j,i}^{f_m}$s' with low values. On the other hand, since $\log(\cdot)$ is an increasing function, an increase in throughput will surely lead to an increase in its log-scaled value. Therefore, maximizing the total log-scaled throughput achieves a good balance between throughput maximization and fairness. Constraints (4.4)–(4.5) in $\mathbf{P}_1$ regulate the usage of the frequency resources at the MeNBs and the SeNBs. Constraint (4.6) enforces a minimum rate requirement for each UE. The data rate for each UE can be calculated as

$$R_k = \sum_{m=1}^{2} \sum_{i=1}^{N_c} \sum_{j=0}^{N_r} x_{k,j,i}^{f_m} n_{k,j,i}^{f_m} R_{k,j,i}^{f_m}. \tag{4.12}$$

4.2 Optimal Resource Allocation Algorithm

It will give the optimal values of α, β and $n_{k,j,i}^{f_m}$ ($m = 1, 2$) by solving $\mathbf{P}_1$. But since $\mathbf{P}_1$ is non-convex, an optimal solution is difficult to obtain. In this section, based on the observation that given α and β values, the optimization problem becomes convex, a two-loop procedure is proposed to solve the optimization problem. The α and β values are optimized in the outer loop using a brute-force search starting from $\alpha(0) = 0$ and $\beta(0) = 0$ and being updated in each step by $\Delta\alpha$ and $\Delta\beta$, respectively. In the inner loop, given each set of α and β values specified in the outer loop, the original optimization problem becomes a constraint convex optimization problem with variables $n_{k,j,i}^{f_m}$. Its optimal value can be found using its dual problem and the Karush-Kuhn-Tucker (KKT) condition for optimality. The detailed optimization procedure is presented as follows.

Introducing Lagrange multipliers $\lambda_{j,i}^{f_m}$, λ_k and $v_k^{f_m}$, the Lagrangian function of the optimization problem $\mathbf{P}_1$ for a given set of α and β values can be formed as

$$
L\left(n_{k,j,i}^{f_m}, \lambda_{j,i}^{f_m}, \lambda_k, v_k^{f_m}\right)
$$

$$
= -\sum_{m=1}^{2} \sum_{i=1}^{N_c} \sum_{j=0}^{N_r} \sum_{k=1}^{N_u} x_{k,j,i}^{f_m} \log(n_{k,j,i}^{f_m} R_{k,j,i}^{f_m}) + \sum_{m=1}^{2} \sum_{i=1}^{N_c} \sum_{j=0}^{N_r} \lambda_{j,i}^{f_m} \left(\sum_{k=1}^{N_u} x_{k,j,i}^{f_m} n_{k,j,i}^{f_m} - f_m \right)
$$

$$
+ \sum_{k=1}^{N_u} \lambda_k \left(R_{\min} - \sum_{m=1}^{2} \sum_{i=1}^{N_c} \sum_{j=0}^{N_r} x_{k,j,i}^{f_m} n_{k,j,i}^{f_m} R_{k,j,i}^{f_m} \right) - \sum_{m=1}^{2} \sum_{i=1}^{N_c} \sum_{j=0}^{N_r} \sum_{k=1}^{N_u} v_{k,j,i}^{f_m} n_{k,j,i}^{f_m}.
$$

$$\tag{4.13}$$

The dual function $g(\lambda_{j,i}^{f_m}, \lambda_k, v_{k,j,i}^{f_m})$ of $\mathbf{P}_1$ is defined as

$$g(\lambda_{j,i}^{fm}, \lambda_k, v_{k,j,i}^{fm}) := \inf_{n_{k,j,i}^{fm}} L(n_{k,j,i}^{fm}, \lambda_{j,i}^{fm}, \lambda_k, v_{k,j,i}^{fm}). \tag{4.14}$$

The dual function is a pointwise minimum of a family of linear functions of the Lagrange multipliers and is always concave regardless of the convexity of the primal function [6]. The optimal value of the original problem is lower bounded by the value of its dual function. The largest lower bound of the primal problem $\mathbf{P}_1$ can be found by solving the following dual problem

$$[\mathbf{P}_2] \quad \max \; g(\lambda_{j,i}^{fm}, \lambda_k, v_{k,j,i}^{fm}) \tag{4.15}$$

$$\text{subject to } \lambda_{j,i}^{fm} \geq 0, \; \lambda_k \geq 0, \; v_{k,j,i}^{fm} \geq 0. \tag{4.16}$$

Due to the concavity of the $g(\lambda_{j,i}^{fm}, \lambda_k, v_{k,j,i}^{fm})$ function, the dual problem $\mathbf{P}_1$ is a convex optimization problem. Since the functions in constraints (4.4)–(4.6) and (4.8) are affine functions of $n_{k,j,i}^{fm}$, the primal problem satisfies the weak form of Slater's condition [7]. Together with the fact that the primal problem is a convex problem given fixed α and β values, it can conclude that strong duality holds, i.e., the gap between the optimal solution of the primal problem and the optimal solution of the dual problem is zero. Therefore, instead of solving the primal problem directly, its dual problem can be solved alternatively.

4.2.1   Optimal $\lambda_{j,i}^{fm}$, λ_k, and $v_{k,j,i}^{fm}$ Values from the Dual Problem

According to [7], an explicit expression of the dual function can be calculated using its conjugate function as

$$g\left(\lambda_{j,i}^{fm}, \lambda_k, v_{k,j,i}^{fm}\right)$$

$$= \inf_{n_{k,j,i}^{fm}} L_{k,j,i}\left(n_{k,j,i}^{fm}, \lambda_{j,i}^{fm}, \lambda_k, v_{k,j,i}^{fm}\right)$$

$$= -\sum_{m=1}^{2}\sum_{i=1}^{N_c}\sum_{j=0}^{N_r}\sum_{k=1}^{N_u} \lambda_{j,i}^{fm} f_m + \sum_{k=1}^{N_u} \lambda_k R_{\min}$$

$$+ \inf_{n_{k,j,i}^{fm}} \sum_{m=1}^{2}\sum_{i=1}^{N_c}\sum_{j=0}^{N_r}\sum_{k=1}^{N_u} \left(\lambda_{j,i}^{fm} x_{k,j,i}^{fm} n_{k,j,i}^{fm} - v_{k,j,i}^{fm} n_{k,j,i}^{fm}\right.$$

$$\left. -\lambda_k x_{k,j,i}^{fm} R_{k,j,i}^{fm} n_{k,j,i}^{fm} - x_{k,j,i}^{fm} \log\left(R_{k,j,i}^{fm} n_{k,j,i}^{fm}\right)\right)$$

$$= - \sum_{m=1}^{2} \sum_{i=1}^{N_c} \sum_{j=0}^{N_r} \sum_{k=1}^{N_u} \lambda_{j,i}^{fm} f_m \quad + \sum_{k=1}^{N_u} \lambda_k R_{\min}$$

$$+ U^* \left(\left[\lambda_{j,i}^{fm} x_{k,j,i}^{fm} - \lambda_k x_{k,j,i}^{fm} R_{k,j,i}^{fm} - v_{k,j,i}^{fm} \right]_{k=1,\cdots,N_u, j=1,\cdots,N_r, i=1,\cdots,N_c}^{m=1,2} \right)$$

$$(4.17)$$

Where $U^*(\cdot)$ is the conjugate function of the objective function $U(\mathbf{n})$ of the primal problem $\mathbf{P}_1$ given in Eq. (4.3), and is defined as

$$U^*(\mathbf{y}) = \sup_{\mathbf{n} \in \mathbf{dom}\ U} \left(\mathbf{y}^T \mathbf{n} - U(\mathbf{n}) \right). \tag{4.18}$$

Where $y = \lambda_{j,i}^{fm} x_{k,j,i}^{fm} - \lambda_k x_{k,j,i}^{fm} R_{k,j,i}^{fm} - v_{k,j,i}^{fm}$ in this case. According to the definition of conjugate function given in (4.18), the conjugate function of $U(n)$ can be expressed as

$$U^*(y) = \sup_{n \in \mathbf{dom}\ U} (yn - x \log(n)). \tag{4.19}$$

To find the n that maximizes $yn - x \log(n)$, the stationarity condition for unconstrained optimization can be used, given as

$$\nabla_n(yn - x \log(n)) = y - \frac{x}{n} = 0. \tag{4.20}$$

Solving (4.20), it has

$$n^* = \frac{x}{y}. \tag{4.21}$$

Substituting (4.21) into (4.19), then

$$U^*(y) = x(1 - \log(x)) + x \log(y). \tag{4.22}$$

Therefore,

$$U^*(\mathbf{y}) = \sum_{m=1}^{2} \sum_{i=1}^{N_c} \sum_{j=1}^{N_r} \sum_{k=1}^{N_u} \left(x_{k,j,i}^{fm} \left(1 - \log(x_{k,j,i}^{fm}) \right) + x_{k,j,i}^{fm} \log \left(y_{k,j,i}^{fm} \right) \right). \tag{4.23}$$

Substituting (4.23) into (4.17), the closed-form expression of the dual function is in the following.

$$g(\lambda_{j,i}^{fm}, \lambda_k, v_{k,j,i}^{fm}) = -\sum_{m=1}^{2}\sum_{i=1}^{N_c}\sum_{j=0}^{N_r}\sum_{k=1}^{N_u} \lambda_{j,i}^{fm} f_m + \sum_{k=1}^{N_u} \lambda_k R_{\min}$$

$$+\sum_{m=1}^{2}\sum_{i=1}^{N_c}\sum_{j=0}^{N_r}\sum_{k=1}^{N_u} \left(x_{k,j,i}^{fm}\left(1 - \log(x_{k,j,i}^{fm})\right) \right.$$

$$\left. +x_{k,j,i}^{fm} \log\left(\lambda_{j,i}^{fm} x_{k,j,i}^{fm} - \lambda_k x_{k,j,i}^{fm} R_{k,j,i}^{fm} - v_{k,j,i}^{fm}\right) \right) \quad (4.24)$$

Since a dual function is a pointwise infimum of a family of linear functions, it is always concave. The maximum value of $g(\lambda_{j,i}^{fm}, \lambda_k, v_{k,j,i}^{fm})$ can be found using the gradient-based method by simultaneously updating $\lambda_{j,i}^{fm}$, λ_k, and $v_{k,j,i}^{fm}$ along the directions

$$\Delta\lambda_{j,i}^{fm} = \sum_{k=1}^{N_u} \frac{\left(x_{k,j,i}^{fm}\right)^2}{\lambda_{j,i}^{fm} x_{k,j,i}^{fm} - \lambda_k x_{k,j,i}^{fm} R_{k,j,i}^{fm} - v_{k,j,i}^{fm}} - f_m, \quad (4.25)$$

$$\Delta\lambda_k = R_{\min} - \sum_{m=1}^{2}\sum_{i=1}^{N_c}\sum_{j=0}^{N_r} \frac{\left(x_{k,j,i}^{fm}\right)^2 R_{k,j,i}^{fm}}{\lambda_{j,i}^{fm} x_{k,j,i}^{fm} - \lambda_k x_{k,j,i}^{fm} R_{k,j,i}^{fm} - v_{k,j,i}^{fm}}, \quad (4.26)$$

and

$$\Delta v_{k,j,i}^{fm} = \frac{x_{k,j,i}^{fm}}{\lambda_{j,i}^{fm} x_{k,j,i}^{fm} - \lambda_k x_{k,j,i}^{fm} R_{k,j,i}^{fm} - v_{k,j,i}^{fm}}, \quad (4.27)$$

respectively. The updating process continues until it converges to the optimal Lagrange multipliers $\lambda_{j,i}^{fm*}$, λ_k^*, and $v_{k,j,i}^{fm*}$, or the boundaries of $\lambda_{j,i}^{fm} \geq 0$, $\lambda_k \geq 0$, and $v_{k,j,i}^{fm} \geq 0$ are reached. This search guarantees the global optimal solution since the dual problem is concave.

4.2.2 Optimal $n_{k,j,i}^{fm}$ Values

The optimal $n_{k,j,i}^{fm*}$ values can be solved using the KKT conditions with the obtained optimal Lagrange multipliers, which are given as

$$\sum_{k=1}^{N_u} x_{k,j,i}^{fm} n_{k,j,i}^{fm*} - f_m \leq 0 \text{ for } i=1,\cdots,N_c, j=0,\cdots,N_r, m=1,2 \quad (4.28)$$

$$R_{\min} - \sum_{m=1}^{2} \sum_{i=1}^{N_c} \sum_{j=0}^{N_r} x_{k,j,i}^{fm} n_{k,j,i}^{fm*} R_{k,j,i}^{fm} \leq 0 \text{ for } k = 1, \cdots, N_u \tag{4.29}$$

$$-n_{k,j,i}^{fm*} \leq 0 \text{ for } i = 1, \cdots, N_c, j = 0, \cdots, N_r, k = 1, \cdots, N_u,$$
$$m = 1, 2 \tag{4.30}$$

$$\lambda_{j,i}^{fm*} \geq 0 \tag{4.31}$$

$$\lambda_k^* \geq 0 \tag{4.32}$$

$$v_{k,j,i}^{fm*} \geq 0 \tag{4.33}$$

$$\lambda_{j,i}^{fm*} \left(\sum_{k=1}^{N_u} x_{k,j,i}^{fm} n_{k,j,i}^{fm*} - f_m \right) = 0 \text{ for } i = 1, \cdots, N_c, j = 0, \cdots, N_r, \quad m = 1, 2$$
$$\tag{4.34}$$

$$\lambda_k^* \left(R_{\min} - \sum_{m=1}^{2} \sum_{i=1}^{N_c} \sum_{j=0}^{N_r} x_{k,j,i}^{fm} n_{k,j,i}^{fm*} R_{k,j,i}^{fm} \right) = 0 \text{ for } k = 1, \cdots, N_u \tag{4.35}$$

$$v_{k,j,i}^{fm*} n_{k,j,i}^{fm*} = 0 \text{ for } i = 1, \cdots, N_c, j = 0, \cdots, N_r, k = 1, \cdots, N_u, \quad m = 1, 2$$
$$\tag{4.36}$$

$$\nabla_{n_{k,j,i}^{fm}} L(n_{k,j,i}^{fm*}, \lambda_{j,i}^{fm*}, \lambda_k^*, v_k^{fm*}) = 0. \text{ for } i = 1, \cdots, N_c, j = 0, \cdots, N_r,$$
$$k = 1, \cdots, N_u, m = 1, 2 \tag{4.37}$$

The conditions (4.28)–(4.30) represent primal feasibility of $n_{k,j,i}^*$. The conditions (4.31)–(4.33) represent dual feasibility. Conditions (4.34)–(4.36) ensure the complementary slackness for the primal and the dual inequality constraint pairs. Condition (4.37) is the stationarity condition. From (4.37), it has

$$\frac{\partial L_{k,j,i}(n_{k,j,i}^{fm}, \lambda_{j,i}^{fm*}, \lambda_k^*, v_{k,j,i}^{fm*})}{\partial n_{k,j,i}^{fm}} = -\frac{x_{k,j,i}^{fm}}{n_{k,j,i}^{fm}} + \lambda_{j,i}^{fm*} x_{k,j,i}^{fm} - \lambda_k^* x_{k,j,i}^{fm} R_{k,j,i}^{fm} - v_{k,j,i}^{fm*} = 0,$$
$$\tag{4.38}$$

obtaining

$$n_{k,j,i}^{fm*} = \frac{x_{k,j,i}^{fm}}{\lambda_{j,i}^{fm*} x_{k,j,i}^{fm} - \lambda_k^* x_{k,j,i}^{fm} R_{k,j,i}^{fm} - v_{k,j,i}^{fm*}}. \tag{4.39}$$

It can be shown that the $\lambda_{j,i}^{fm*}$, λ_k^*, $v_{k,j,i}^{fm*}$ and $n_{k,j,i}^{fm*}$ values calculated using the proposed approach satisfy all the conditions (4.28)–(4.37). Since for convex optimization problems, satisfying the KKT conditions ensures global optimality, the $n_{k,j,i}^{fm*}$ values given in (4.39) are the optimal solution of the problem for the given α and β. The detailed proof on optimality by KKT Condition is in the following.

Substituting $n_{k,j,i}^{fm*}$ given in (4.39) into (4.34) and (4.35), the two conditions become

$$\lambda_{j,i}^{fm*} \left(\sum_{k=1}^{N_u} \frac{\left(x_{k,j,i}^{fm}\right)^2}{\lambda_{j,i}^{fm} x_{k,j,i}^{fm} - \lambda_k x_{k,j,i}^{fm} R_{k,j,i}^{fm} - \nu_{k,j,i}^{fm}} - f_m \right) = 0, \qquad (4.40)$$

and

$$\lambda_k^* \left(R_{\min} - \sum_{m=1}^{2} \sum_{i=1}^{N_c} \sum_{j=0}^{N_r} \frac{\left(x_{k,j,i}^{fm}\right)^2 R_{k,j,i}^{fm}}{\lambda_{j,i}^{fm} x_{k,j,i}^{fm} - \lambda_k x_{k,j,i}^{fm} R_{k,j,i}^{fm} - \nu_{k,j,i}^{fm}} \right) = 0. \qquad (4.41)$$

From the previous discussion in solving the dual problem to calculate $\lambda_{j,i}^{fm*}$ and λ_k^*, when $\lambda_{k,j,i}^{fm} = \lambda_{k,j,i}^{fm*}$ and $\lambda_k = \lambda_k^*$, then $\Delta\lambda_{j,i}^{fm*} = 0$ and $\Delta\lambda_k^* = 0$. By comparing the $\Delta\lambda_{j,i}^{fm}$ and $\Delta\lambda_k$ expressions given in (4.25) and (4.26) with the two conditions in (4.40) and (4.41), it can be seen that the two KKT conditions can be satisfied by the optimal $n_{k,j,i}^{fm*}$, $\lambda_{j,i}^{fm*}$ and λ_k^* values. Moreover, by $\Delta\lambda_{j,i}^{fm*} = 0$ and $\Delta\lambda_k^* = 0$ given in (4.25) and (4.26), the KKT conditions (4.28) and (4.29) are satisfied with equality. From the calculation of the dual function, the value $\lambda_{j,i}^{fm} x_{k,j,i}^{fm} - \lambda_k x_{k,j,i}^{fm} R_{k,j,i}^{fm} - \nu_{k,j,i}^{fm}$ appears in the domain of the log function. This implies that $\lambda_{j,i}^{fm} x_{k,j,i}^{fm} - \lambda_k x_{k,j,i}^{fm} R_{k,j,i}^{fm} - \nu_{k,j,i}^{fm} > 0$, and consequently $n_{k,j,i}^{fm} \geq 0$. The KKT condition (4.30) is satisfied. From the $\Delta\nu_{k,j,i}^{fm}$ expression given in (4.27), $\Delta\nu_{k,j,i}^{fm}$ could be positive or negative but could never be zero if $x_{k,j,i}^{fm} \neq 0$. This indicates that the optimal value of $\nu_{k,j,i}^{fm}$ is located at its boundary, i.e., $\nu_{k,j,i}^{fm} = 0$. The KKT condition (4.36) is satisfied. Since for each given set of α and β values, the optimization problem $\mathbf{P}_1$ is a convex optimization problem, satisfying the KKT condition is necessary and sufficient for optimality. The $n_{k,j,i}^{fm*}$ value given in (4.39) is the optimal solution of the problem.

4.2.3   Summary of the Two-Loop Optimization Algorithm

For each pair of α and β values given in the outer loop, the optimal value of the objective function (4.3) can be calculated in the inner loop using the above-mentioned optimization process. All the optimal values achieved at different (α, β) pairs then can be compared, and the maximum one can be selected. The corresponding α^*, β^* and $n_{k,j,i}^{fm*}$ values are the optimal solution of problem $\mathbf{P}_1$. A summary of the proposed two-loop optimization procedure is given as follows.

Step 1: Outer loop initialization. Set $\tau_\alpha = 0$ and $\tau_\beta = 0$, initialize $\alpha(0) = 0$ and $\beta(0) = 0$. Choose the update steps, $\Delta\alpha$ and $\Delta\beta$, for α and β, respectively.

Step 2: Inner loop iteration.

1. Initialize $\lambda_{j,i}^{fm}(0)$, $\lambda_k(0)$, and $v_{k,j,i}^{fm}(0)$. Update subband association $x_{k,j,i}^{fm}$ based on the β value specified in the outer loop.
2. At the tth inner loop iteration, compute $\Delta\lambda_{j,i}^{fm}(t)$, $\Delta\lambda_k(t)$, and $\Delta v_{k,j,i}^{fm}(t)$ using (4.25)–(4.27). Update $\lambda_{j,i}^{fm}(t)$, $\lambda_k(t)$, and $v_{k,j,i}^{fm}(t)$ as

$$\lambda_{j,i}^{fm}(t) = \lambda_{j,i}^{fm}(t-1) + \mu\Delta\lambda_{j,i}^{fm}(t), \tag{4.42}$$

$$\lambda_k(t) = \lambda_k(t-1) + \mu\Delta\lambda_k(t), \tag{4.43}$$

and

$$v_{k,j,i}^{fm}(t) = \lambda_{k,j,i}^{fm}(t-1) + \mu\Delta v_{k,j,i}^{fm}(t), \tag{4.44}$$

where μ is the step size of each update. If $|\Delta\lambda_{j,i}^{fm}(t)| \leq \epsilon$, or $|\Delta\lambda_k(t)| \leq \epsilon$, or $v_{k,j,i}^{fm}(t) \leq \epsilon$, then $\lambda_{j,i}^{fm*} = \lambda_{j,i}^{fm}(t)$, or $\lambda_k^{fm*} = \lambda_k^{fm}(t)$, or $v_{k,j,i}^{fm*} = v_{k,j,i}^{fm}(t)$. When all the $\lambda_{j,i}^{fm*}$, λ_k^* and $v_{k,j,i}^{fm*}$ are reached, stop inner loop updating. Otherwise, let $t = t + 1$ and go to **Step 2**(2) to keep on updating in the inner loop.

3. Substitute the obtained $\lambda_{j,i}^{fm*}$, λ_k^* and $v_{k,j,i}^{fm*}$ values into (4.24) to compute the optimal value of the dual function $g(\alpha(\tau_\alpha), \beta(\tau_\beta), \lambda_{j,i}^{fm*}, \lambda_k^*, v_{k,j,i}^{fm*})$, which is also the optimal value of the objective function in $\mathbf{P}_1$ for the $\alpha(\tau_\alpha)$ and $\beta(\tau_\beta)$ values given in the outer loop.

Step 3: Outer loop update for α. Set $\tau_\alpha = \tau_\alpha + 1$, update $\alpha(\tau_\alpha)$ value as

$$\alpha(\tau_\alpha) = \alpha(\tau_\alpha - 1) + \Delta\alpha. \tag{4.45}$$

If $\alpha(\tau_\alpha) \in [0, 1]$, then go back to **Step 2** to start a new cycle of inner loop iteration. Otherwise, go to **Step 4**.

Step 4: Outer loop update for β. Set $\tau_\beta = \tau_\beta + 1$, update $\beta(\tau_\beta)$ value as

$$\beta(\tau_\beta) = \beta(\tau_\beta - 1) + \Delta\beta. \tag{4.46}$$

If $\beta(\tau_\beta) \in [0, 1]$, then go back to **Step 2**. Otherwise, go to **Step 5**.

Step 5: Find the global optimal solution. Among all the $(\alpha(\tau), \beta(\tau), \lambda_{j,i}^{fm*}, \lambda_k^{fm*}, v_{k,j,i}^{fm*})$ sets, the one that gives the largest $g(\alpha(\tau), \beta(\tau), \lambda_{j,i}^{fm*}, \lambda_k^{fm*}, v_{k,j,i}^{fm*})$ value gives the optimal solution of the optimal problem $\mathbf{P}_1$. Substituting the corresponding $\lambda_{j,i}^{fm*}$, λ_k^{fm*}, and $v_{k,j,i}^{fm*}$ values into (4.39), we can obtain the optimal $n_{k,j,i}^{fm*}$ values.

Since the dual function $g(\lambda_{j,k}^{fm}, \lambda_k, v_{k,j,i}^{fm})$ is concave, using gradient-descent method, the optimal values of $\lambda_{j,k}^{fm}, \lambda_k, v_{k,j,i}^{fm}$ can be found in linear time. According to [7], the number of iterations is bounded by

$$K_\lambda = \frac{\max_{\lambda_{j,i}^{fm}, \lambda_k, v_{k,j,i}^{fm}} \log\left(\frac{g(\lambda_{j,i}^{fm}, \lambda_k, v_{k,j,i}^{fm}) - g(\lambda_{j,i}^{fm*}, \lambda_k^*, v_{k,j,i}^{fm*})}{\epsilon}\right)}{\log(1/c)}, \tag{4.47}$$

where $c = 1 - m_n/M_n$, $m_n = \inf_{\lambda_{j,i}^{fm}, \lambda_k, v_{k,j,i}^{fm}} \nabla^2 g(\lambda_{j,i}^{fm}, \lambda_k, v_{k,j,i}^{fm})$, $M_n = \sup_{\lambda_{j,i}^{fm}, \lambda_k, v_{k,j,i}^{fm}} \nabla^2 g(\lambda_{j,i}^{fm}, \lambda_k, v_{k,j,i}^{fm})$, and ϵ is a small value regulating the maximum gap between the optimization solution and the optimal value. In the outer loop, the number of iterations for updating α and β values are $K_\alpha = 1/\Delta\alpha$ and $K_\beta = 1/\Delta\beta$, respectively. The total computation complexity in terms of number of iterations is thus upper bounded by $K_\lambda K_\alpha K_\beta$.

Note that the use of Lagrangian dual decomposes the inner optimization problem of optimizing $n_{k,j,i}^{fm}$ into a set of parallel sub-problems that can be optimized independently at each macro and small nodes. For the outer optimization on α and β, however, a centralized optimization considering the overall long-term radio resource allocation is required. Nevertheless, such a centralized optimization would not complicate implementation as the optimization is based on the long-term resource allocation $n_{k,j,i}^{fm}$ and can be done pseudo-statically in the network central controller. Once determining the optimal α and β values, the system only needs to find the instantaneous resource allocation $n_{k,j,i}^{fm}(t)$ values that fit into the instantaneous channel state in the tth time instance. Optimization on $n_{k,j,i}^{fm}(t)$ can still use the Lagrangian dual decomposition based approach and thus can be implemented distributively. The long-term average of $n_{k,j,i}^{fm}(t)$ should approach $n_{k,j,i}^{fm}$.

4.3 Performance Results and Discussion

The performance of the optimal resource allocation scheme is simulated in a LTE heterogeneous network with a 19-cell 3-sector three-ring hexagonal cell structure with a cell radius at 2 km. Four SeNBs are uniformly deployed in each sector. Simulation setup follows the guidelines described in 3GPP technical reports [8]. Transmit power of the MeNB is 46 dBm (40 W) and transmit power of the SeNB is 30 dBm (1 W). UEs are uniformly distributed in the network with an average of 200 UEs in each cell. Each UE requires a minimum data rate of $R_{\min} = 8.6$ kbps.

The simulation shows that the global optimality of the objective function is achieved at $(\alpha^*, \beta^*) = (0.17, 0.7)$. This result indicates that it is optimal to allocate f_1 sub-band with about $2/3$ of the total frequency resource and let the MeNBs

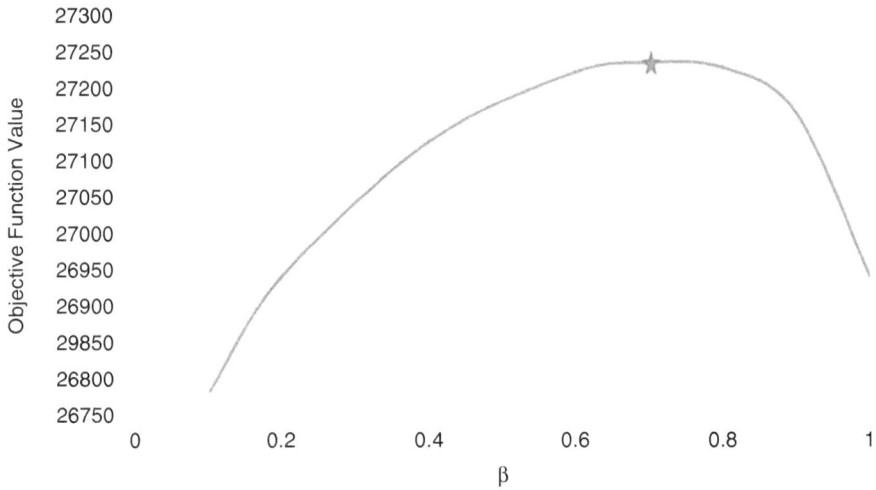

Fig. 4.2 Objective function value for fixed $\alpha = 0.17$, different β

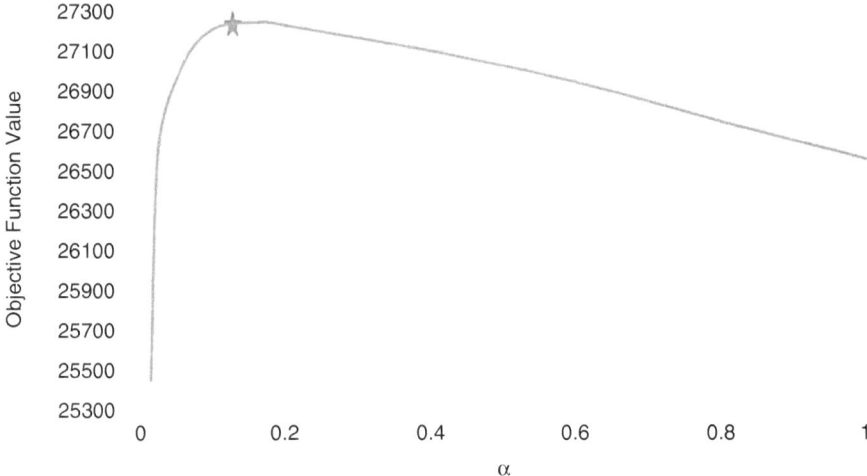

Fig. 4.3 Objective function value for fixed $\beta = 0.7$, different α

transmit at a power of 6.8 W at its inner cell, in order to achieve the maximum system efficiency. By using this FFR and power control scheme, the UEs in the outer range of the small cells receive low interference, while the UEs in the inner range of the MeNB cell could still achieve a relatively satisfactory rate. Both spectrum efficiency and user experience can be achieved. Figures 4.2 and 4.3 show one of the values of α and β is fixed and the objective function value is plotted with respect to

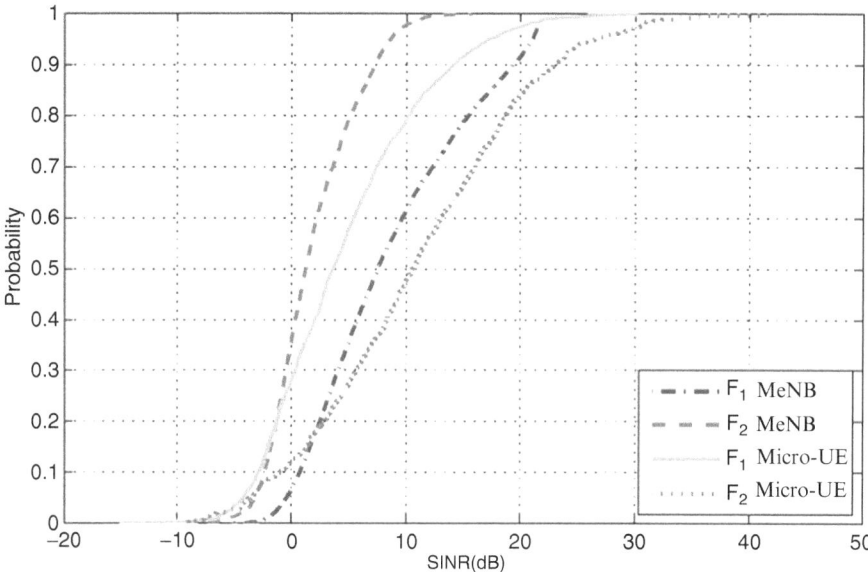

Fig. 4.4 Empirical cumulative distribution for SINR of UEs under the optimal α and β values, $\alpha = 0.17$ and $\beta = 0.7$

the different values of the other variable, to get a better idea on the variation of the objective function with respect to the α and β values. Note that $\alpha = 1$ and $\beta = 0$ corresponds to the case with no FFR and no power control. By comparing the cases with optimal α and β values, it can be seen the advantages of the proposed FFR and resource allocation scheme in improving network performance.

In Fig. 4.4, the SINR curves of the above simulated case under the optimal α and β values are plotted. As a comparison, the (α, β) value is changed to be $(1, 0)$ and the SINR curves are plotted in Fig. 4.5. Note that when setting $(\alpha, \beta) = (1, 0)$, the system reduces to a no FFR and no power control system. By comparing the SINR curves of the two cases, it can be seen that by using the optimal setting, the SINR distribution of the UEs in the outer range of the small cell (f_1 SeNB) is greatly improved. Meanwhile, a slight degradation of the SINR distribution of the UEs in the inner range of the MeNB cell (f_1 MeNB) can be observed. This result indicates that without power control at the MeNB, the UEs in the outer region of the SeNB cell suffer from high interference and receive low data rate. With power control and fairness consideration, the SINR of the UEs in the outer region of the SeNB cell can be greatly improved. On the other hand, since the UEs in the inner region of the MeNB cell are the most advantageous UEs, i.e., possess good channel quality and enjoy low interference, reducing transmit power to these UEs does not affect much on the performance.

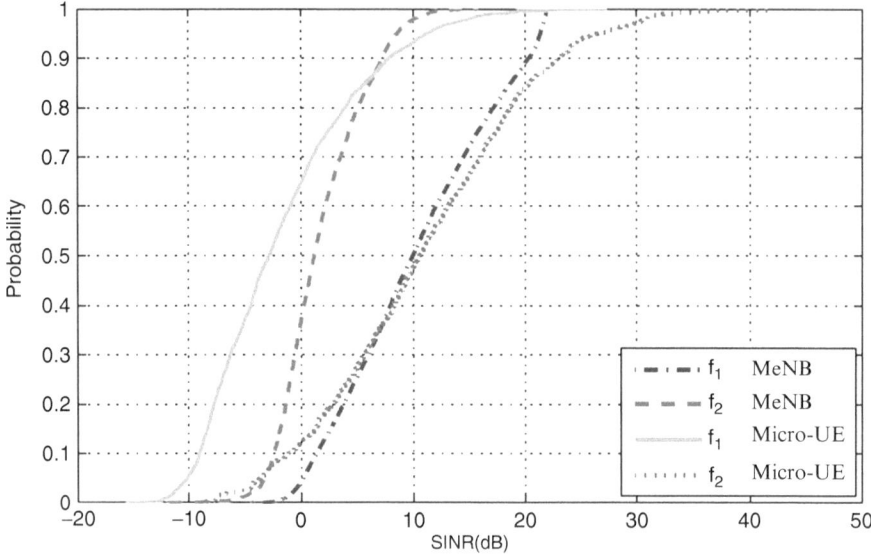

Fig. 4.5 Empirical cumulative distribution for SINR of UEs under $\alpha = 1$ and $\beta = 0$

4.4   Summary

This chapter studied an optimal downlink radio resource management scheme in an LTE heterogeneous network. A fractional frequency reuse and power control scheme is illustrated that can coordinate the interference between MeNBs and SeNBs. An optimal algorithm that jointly optimizes the fractional frequency reuse parameters and the frequency resource allocation among the UEs is presented with the objective of achieving both spectrum efficiency and user fairness. A two-loop optimization algorithm, where a closed-form solution of the resource allocation can be derived using dual problem and KKT condition, is presented for solving the problem. Simulation results show that the network performance can be greatly improved by the fractional frequency reuse scheme and the optimization framework.

References

1. R. Q. Hu, Y. Qian, and W. Li, "On the downlink time, frequency and power coordination in an LTE relay network," in *Proc. of IEEE GLOBECOM 2011*, Houston, Texas, Dec. 2011.
2. Q. Li, R. Q. Hu, Y. Xu, and Y. Qian, "Optimal Fractional Frequency Reuse and Power Control in the Heterogeneous Wireless Networks," *IEEE Transactions on Wireless Communications*, Vol. 12, No. 6, pp. 2658–2668, Jun. 2013.
3. L. Tassiulas and S. Sarkar, "Maxmin fair scheduling in wireless networks," in *Proc. of IEEE INFOCOM 2002*, New York, NY, pp. 763–772, Jun. 2002.

4. P. Viswanath, D. Tse, and R. Laroia, "Opportunistic beamforming using dumb antennas," *IEEE Trans. Inform. Theory*, vol. 48, pp. 1277–1294, Jun. 2002.

5. J. Sun, E. Modiano, and L. Zheng, "Wireless channel allocation using an auction algorithm," *IEEE J. Select. Areas Commun.*, vol. 24, pp. 1085–1096, May 2006.

6. Z. Q. Luo, W. Yu, "An introduction to convex optimization for communications and signal processing," *IEEE J. Select. Areas Commun.*, vol. 24, no. 8, pp. 1426–1438, Aug. 2006.

7. S. Boyd and V. Vandenberghe, "Convex Optimization," Cambridge University Press, 2004.

8. 3GPP TR36.814, "Further advancements for E-UTRA physical layer aspects," v9.0.0, Mar. 2010.

Chapter 5
Radio Resource Allocation in Heterogeneous Networks

5.1 Intra-cell CoMP Scheme and Communication Model

Intra-cell coordinated multiple point (CoMP) transmission is discussed in Chap. 2. Downlink intra-cell cooperative transmission and optimal intra-cell CoMP resource allocation schemes are investigated in heterogeneous networks with cooperative relays in [1–3]. In this chapter, radio resource allocation schemes for two-tier heterogeneous networks in LTE are further discussed. Radio resource allocation schemes with intra-cell CoMP and in-band wireless backhaul are studied, and an optimal framework with resource allocation strategy is presented that is asymptotically optimal on the proportional fairness metric. Figure 5.1 demonstrates the considered intra-cell CoMP scenario. Depending on the received SINR, UEs in the coverage of the small cells can be served either solely by the RN or jointly by the donor MeNB and the RN. The UEs at the boundary of the small cell may receive a relatively low SINR from the RN and a high SINR from the MeNB and are thus the greatest beneficiary from the CoMP transmissions. Upon receiving the joint signals from the MeNB and the RN, the UE extracts its information using maximum likelihood decoding or other suboptimal decoding methods.

The total frequency resources are divided into F resource blocks (RB) and OFDMA is the downlink physical layer transmission scheme. UEs will be assigned with an integer number of RBs. The assignment is determined by the scheduling and changes from one subframe to another. Assume that the wireless channel is frequency-selective across RBs and frequency-flat within each RB. Denote the frequency-domain channel gain on the fth RB at time t between the ith MeNB and the kth UE as $h_{k,0,i}^{f}(t)$, and between the jth RN in the ith sector and the kth UE as $h_{k,j,i}^{f}(t)$. The channel gain counts both long-term path loss and shadowing and short-term fading due to multipath and mobility. The received SINR of the M-UEs, R-UEs can be calculated from (2.1) and (2.2). The received SINR of the C-UEs at time t can be evaluated respectively as

R.Q. Hu and Y. Qian, *Resource Management for Heterogeneous Networks in LTE Systems*, SpringerBriefs in Electrical and Computer Engineering, DOI 10.1007/978-1-4939-0372-6_5, © The Author(s) 2014

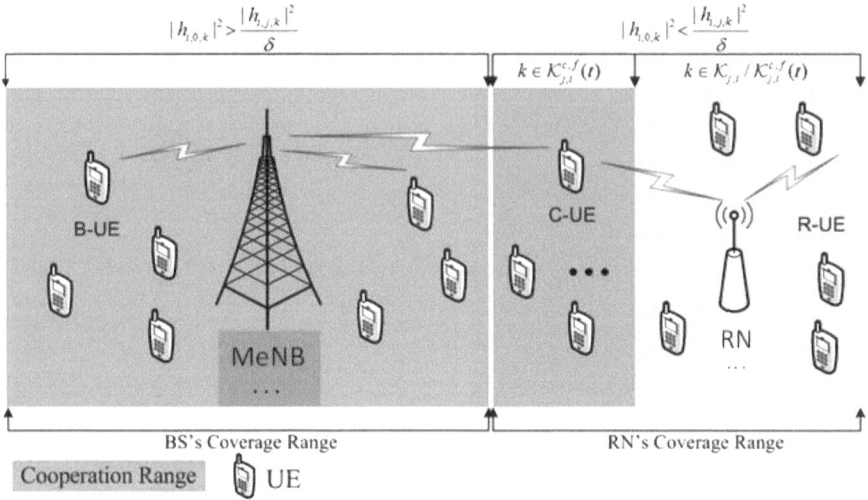

Fig. 5.1 A heterogeneous network with RNs and intra-cell CoMP

$$\text{SINR}_{k,j,i}^{c,f}(t) = \frac{P_m|h_{k,0,i}^f(t)|^2 + P_p|h_{k,j,i}^f(t)|^2}{\sum_{\substack{i'=1\\i'\neq i}}^{N_c}|h_{k,0,i'}^f(t)|^2 P_m + \sum_{i=1}^{N_c}\sum_{\substack{j'=1\\j'\neq j}}^{N_r}|h_{k,j',i}^f(t)|^2 P_r + N_0}. \quad (5.1)$$

The data rate in terms of bit/s/Hz for the kth C-UE on the fth RB at time t can be calculated using Shannon formula as

$$R_{k,j,i}^{c,f}(t) = \log\left(1 + \text{SINR}_{k,j,i}^{c,f}(t)\right). \quad (5.2)$$

For RNs with wireless in-band backhaul connection with the MeNBs, the backhaul link between the donor MeNB and the RN uses the same frequency band as the direct link and the access link and the resource sharing is done in a TDD mode, as shown in Fig. 5.2, where over an interval of T subframes, the backhaul communication is allocated with T_1 subframes and the direct link and access link communications are allocated with $T_2 = T - T_1$ subframes. A scheduling scheme for such a TDD-based communication is demonstrated in Fig. 5.3, where a subframe is the scheduling unit considered in this paper. Communication in the backhaul link takes place in one subframe every T_b subframes, where T_b is a system design parameter to be optimized. During the backhaul transmission subframes, RNs receive from their donor MeNBs. The RNs store the received information in their buffers and subsequently forward the information to the corresponding UEs in the appropriate time/frequency resources. Assume high-capacity and constant-quality backhaul links, this assumption can be justified by the fact that RNs are usually equipped with multiple antennas and placed in locations with low shadowing. Given

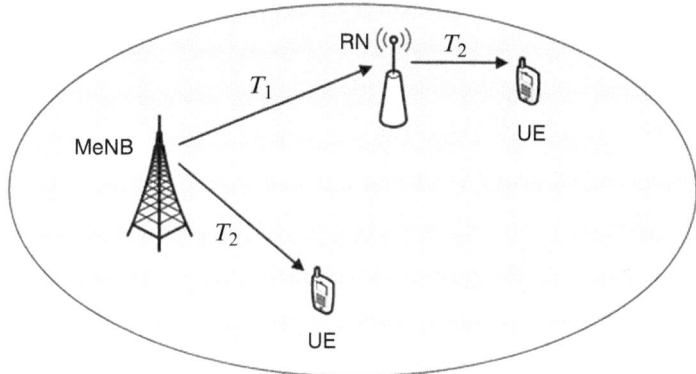

Fig. 5.2 TDD transmission mode of the direct/access links and the backhaul links

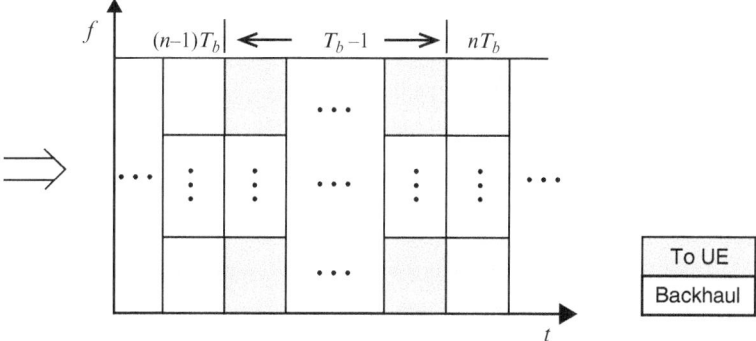

Fig. 5.3 Practical scheduling for the time-duplex transmission of the direct/access links and the backhaul links

the data rate definition in (5.2), to support the kth R-UE transmission in a RN's RB, the frequency resource required in the backhaul link is

$$a_{k,j,i}^{f}(t) = \frac{R_{k,j,i}^{f}(t)}{R_{0,j,i}^{f}}.$$ (5.3)

Define $a_{k,j,i}^{f}(t)$ as the ratio of backhaul utility. For high-capacity backhauls, $a_{k,j,i}^{f}(t) < 1$. The ratio of backhaul utility for the kth C-UEs can be similarly calculated as

$$a_{k,j,i}^{c,f}(t) = \frac{R_{k,j,i}^{c,f}(t)}{R_{0,j,i}^{f}}.$$ (5.4)

5.2   The Optimal Resource Allocation Framework

The objective is to optimize network long-term spectrum efficiency and to ensure fairness among the UEs. To this end, it should (1) properly associate each UE with a MeNB or a RN, (2) for the UEs associated with the RNs, decide whether to use CoMP or not, and (3) properly allocate the frequency resources to the UEs during each scheduling interval.

For mobile association, either range-expansion based or load-balancing based mobile association can be used. Here, as the main focus of this chapter is on radio resource allocation, range-expansion based mobile association is used for illustration purpose. The parameters $x_{k,0,i}$ and $x_{k,j,i}$ as defined in Chap. 3 are again used in indicating the UE association status with the MeNB and the RN. For the R-UEs, it needs to further decide whether CoMP is used or not. An SINR threshold σ and an interference threshold $\theta \in [0, 1]$ both need to be determined, such that if the R-UE receives an SINR from its associated RN at a level below σ and a signal power from the donor MeNB at a level higher than θ times of its total received interference, the UE is a C-UE and is cooperatively served by its anchor RN and the donor MeNB. The thresholds σ and θ select the UEs that are located at the edge of the RN cell and can benefit most from the cooperative transmission. CoMP decision needs to be made for each RB. More specifically, on the fth RB, the set of C-UEs $\mathcal{K}_{j,i}^{c,f}(t)$ associated with the jth RN in the ith can be established as

$$\mathcal{K}_{j,i}^{c,f}(t) = \{k \in \mathcal{K}_{j,i} | \text{SINR}_{k,j,i}^{f}(t) < \sigma \text{ and } |h_{k,0,i}^{f}(t)|^2 P_b > \theta I_k^{f}(t)\}, \quad (5.5)$$

where $I_k^{f}(t)$ is the total interference received by the kth UE and $\mathcal{K}_{j,i}$ is the set of R-UEs associated with the jth RN in the ith macro cell sector. As demonstrated in Fig. 5.1, the whole communication area is either covered by the MeNB or the RN. The boundary delineates the coverage areas of MeNB and RN. Within the coverage range of the RN, a set of C-UEs is selected according to (5.5). Denote $\mathcal{K}_{0,i}$ as the set of M-UEs associated with the ith MeNB and $\mathcal{J}_i$ as the set of RNs in the ith sector.

The parameter $x_{k,j,i}^{c,f}(t)$ is used to indicate if CoMP is used or not. $x_{k,j,i}^{c,f}(t) = 1$ indicates that CoMP is applied for UE k associated with RN j in sector i while $x_{k,j,i}^{c,f}(t) = 0$ indicates otherwise. $x_{k,j,i}^{c,f}(t)$ is a function of t since it will be decided at the beginning of each scheduling interval while $x_{k,j,i}$ and $x_{k,0,i}$ are decided once the association decision is made. Assume that each UE in the system can be associated with a MeNB or a RN or associated with no node, i.e., not currently served by any node. Then $x_{k,j,i}$ satisfies

$$\sum_{i=1}^{N_c} \sum_{j=0}^{N_r} x_{k,j,i} \leq 1, \; \forall k. \quad (5.6)$$

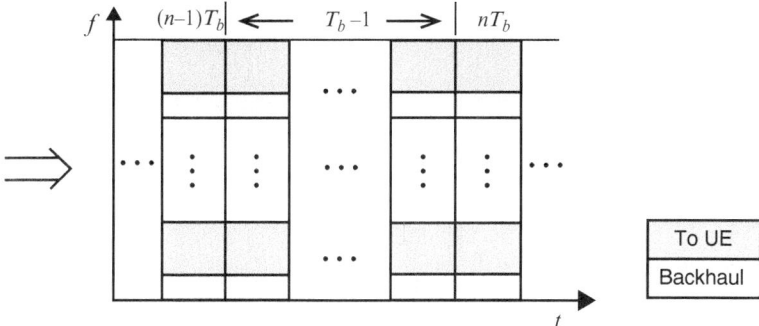

Fig. 5.4 Mathematical equivalence for the time-duplex transmission of the direct/access links and the backhaul links

For all the UEs that are granted into the network, scheduling assigns proper radio resources at each subframe to these UEs. For systems with RNs that support wireless in-band backhauls, radio resource management needs to take into account resources consumed on the backhaul links as well as on the direct and access links. The resource allocations on the direct links, access links and backhaul links are integral parts of a global system optimization problem, which will need to decide the optimal T_b value and optimal allocation for each RB on the access and direct links. To facilitate mathematical formulation and to gain more insights on the impact of backhaul resource consumption on the system RRM, instead of counting the backhaul resource consumption in a subframe basis as shown in Fig. 5.3, the backhaul resource consumption is distributed into each RB as shown in Fig. 5.4. Each MeNB's RB can be considered to be virtually shared by downlink transmission towards one of its associated M-UEs or C-UEs on the direct link and by backhaul transmission towards its associated RNs, and each of RN's RB is considered to be virtually shared by the reception from its donor MeNB on the backhaul link and downlink transmission towards its associated R-UEs or C-UEs on the access link. Backhaul transmission is only needed when there are UEs associated with RNs. The radio resource spent on the backhaul link can be determined by $a_{k,j,i}^{f}(t)$ and $a_{k,j,i}^{c,f}(t)$ values as defined in (5.3) and (5.4). The scheduling will ultimately decide (1) the UE assigned for each RB of each MeNB/RN, and (2) the T_b value that schedules direct/access link transmission and backhaul link transmission. The following variables are further introduced to formulate the scheduling problem. Denote $n_{k,0,i}^{f}(t)$ as the portion of the fth RB assigned to the kth UE at the ith MeNB in the tth subframe, $n_{k,j,i}^{f}(t)$ as the portion of the fth RB assigned to the kth UE at the jth RN in the ith sector and tth subframe, $n_{k,j,i}^{c,f}(t)$ as the portion of the fth RB assigned to the kth C-UE associated with the jth RN in the ith sector and the tth subframe, and $n_{k,j,i}^{b,f}(t)$ as the portion of the fth RB assigned to the backhaul link between the jth RN and the ith MeNB in support of the kth UE. $n_{k,j,i}^{b,f}(t)$ can be calculated as

$$n_{k,j,i}^{b,f}(t) = a_{k,j,i}^{f}(t)n_{k,j,i}^{f}(t), \tag{5.7}$$

for the B-UEs/R-UEs and

$$n_{k,j,i}^{b,f}(t) = a_{k,j,i}^{c,f}(t)n_{k,j,i}^{c,f}(t), \tag{5.8}$$

for the C-UEs. The $n_{k,j,i}^{c,f}(t)$ portion of RBs is required at both the i MeNB and the jth RN to support the kth C-UE. By the above definitions of $n_{k,j,i}^{f}(t)$ and $n_{k,j,i}^{c,f}(t)$, a RB is allowed to be shared by multiple UEs. In the next section, it can be shown that in the asymptotically optimal solution, for each RB of the ith MeNB or the jth RN in the ith MeNB, only one of the $n_{k,j,i}^{f}(t)$s' or $n_{k,j,i}^{c,f}(t)$s' is nonzero, meaning that a single RB will be assigned to only one UE. This result is in consistent with most rules used in practical implementation.

Proportional fairness is used as the performance metric to ensure a good tradeoff between spectrum efficiency and fairness. The optimization problem with a long-term proportional fair resource allocation is thus formulated as

$$[\mathbf{P}_1] \quad \max \sum_k \log(R_k(t)) \tag{5.9}$$

subject to

$$\sum_{k=1}^{N_u} x_{k,0,i} n_{k,0,i}^{f}(t) + \sum_{j=1}^{N_r}\sum_{k=1}^{N_u} x_{k,j,i} x_{k,j,i}^{c,f}(t) n_{k,j,i}^{c,f}(t) + \sum_{j=1}^{N_r}\sum_{k=1}^{N_u} x_{k,j,i} n_{k,j,i}^{b,f}(t) \leq 1$$

$$\text{for } i = 1, \cdots, N_c, f = 1, \cdots, F \tag{5.10}$$

$$\sum_{k=1}^{N_u} x_{k,j,i}(1 - x_{k,j,i}^{c,f}(t)) n_{k,j,i}^{f}(t) + \sum_{k=1}^{N_u} x_{k,j,i} x_{k,j,i}^{c,f}(t) n_{k,j,i}^{c,f}(t)$$

$$+ \sum_{k=1}^{N_u} x_{k,j,i} n_{k,j,i}^{b,f}(t) \leq 1 \quad \text{for } i = 1, \cdots, N_c, \ j = 1, \cdots, N_r, f = 1, \cdots, F \tag{5.11}$$

$$n_{k,j,i}^{f}(t) \geq 0 \ \forall i, j, k, f \tag{5.12}$$

$$n_{k,j,i}^{c,f}(t) \geq 0 \ \forall i, j, k, f, \tag{5.13}$$

where

$$R_k(t) = \frac{1}{T_c} \sum_{\tau=t-T_c+1}^{t} S_k(\tau), \qquad (5.14)$$

T_c is the size of the time window for moving average, and $S_k(\tau)$ is the moving average system throughput, which is expressed as:

$$S_k(\tau) = \sum_{f=1}^{F} \sum_{i=1}^{N_c} \sum_{j=0}^{N_r} \left(x_{k,j,i} (1 - x_{k,j,i}^{c,f}(t)) R_{k,j,i}^{f}(\tau) n_{k,j,i}^{f}(\tau) \right.$$

$$\left. + x_{k,j,i} x_{k,j,i}^{c,f}(\tau) R_{k,j,i}^{c,f}(\tau) n_{k,j,i}^{c,f}(\tau) \right), \qquad (5.15)$$

where let $x_{k,0,i}^{c,f}(t) = 0$ for notational consistency.

By solving the $n_{k,j,i}^{f}(t)$ and $n_{k,j,i}^{c,f}(t)$ values in $\mathbf{P}_1$, it can find the allocated UE for each RB and the T_b value for scheduling the direct/access link transmission and the backhaul link transmission. Constraint (5.10) is the resource constraint for each RN at the MeNB. The first term in (5.10) computes the portion of RB f used by the direct link, the second term in (5.10) computes the portion of the RB f used by serving the C-UEs, and the third term in (5.10) computes the portion of RB f used by the backhaul link. Constraint (5.11) gives the resource constraint for each RN at the MeNB. The first and second terms in (5.11) calculate the portion of RB f used by the R-UEs and C-UEs, respectively. The third term in (5.11) calculates the portion of RB f used by the backhaul link.

As a multicarrier proportional fair scheduling problem, the computational complexity in finding the optimal solution of $\mathbf{P}_1$ is prohibitively high [4]. To fit it for practical implementation, one can apply the gradient-based scheduling algorithm as proposed in [5,6]. It was proven in [6] that the gradient-based scheduling algorithm asymptotically converges to the optimal solution. In the next section, based on the gradient-based scheduling algorithm, it can be shown how to optimally allocate resources in such a heterogeneous network with in-band backhaul RNs.

5.3   An Asymptotically Optimal Radio Resource Allocation Scheme

Using the gradient-based scheduling framework, the system parameters are chosen to maximize the drift of the objective function at each subframe, given as

$$
U(\mathbf{R}(t+1)) - U(\mathbf{R}(t)) = \sum_{k=1}^{N_u} \left(\log \left(R_k(t) + \epsilon \big(S_k(t+1) - S_k(t - T_c + 1)\big) \right) \right.
$$

$$
\left. - \log \big(R_k(t) \big) \right)
$$

$$
= \sum_{k=1}^{N_u} \frac{1}{R_k(t)} S_k(t+1)\epsilon - \sum_{k=1}^{N_u} \frac{1}{R_k(t)} S_k(t - T_c + 1)\epsilon + O(\epsilon^2),
$$

$$(5.16)$$

where $\epsilon = 1/T_c$ and the second equality is obtained using first order Taylor expansion. Since only the first term in (5.16) depends on future decisions, the gradient-based scheduling problem can be formulated as

$$
[\mathbf{P_2}] \quad \max_{n_{k,j,i}^{f}(t), n_{k,j,i}^{c,f}(t)} \sum_{k=1}^{N_u} \frac{1}{R_k(t-1)} \times \sum_{f=1}^{F} \sum_{i=1}^{N_c} \sum_{j=0}^{N_r} \left(x_{k,j,i}(1 - - x_{k,j,i}^{c,f}(t)) R_{k,j,i}^{f}(t) n_{k,j,i}^{f}(t) \right.
$$

$$
\left. + x_{k,j,i} x_{k,j,i}^{c,f}(t) R_{k,j,i}^{c,f}(t) n_{k,j,i}^{c,f}(t) \right),
$$

$$(5.17)$$

subject to the constraints given in (5.10)–(5.13). By independence of summation and since the constraints in (5.10)–(5.13) are set on a per RB basis, the optimal solution of $\mathbf{P_2}$ can be found by solving the optimal $n_{k,j,i}^{f}(t)$ and $n_{k,j,i}^{c,f}(t)$ values for each RB using the following optimization formulation.

$$
[\mathbf{P_3}] \quad \max_{n_{k,j,i}^{f}(t), n_{k,j,i}^{c,f}(t)} \sum_{k=1}^{N_u} \sum_{i=1}^{N_c} \sum_{j=0}^{N_r} \frac{1}{R_k(t-1)} \times \left(x_{k,j,i}(1 - x_{k,j,i}^{c,f}(t)) R_{k,j,i}^{f}(t) n_{k,j,i}^{f}(t) \right.
$$

$$
\left. + x_{k,j,i} x_{k,j,i}^{c,f}(t) R_{k,j,i}^{c,f}(t) n_{k,j,i}^{c,f}(t) \right),
$$

$$(5.18)$$

subject to (5.10)–(5.13). By gradient-based scheduling, multi-carrier proportional fair scheduling $\mathbf{P_1}$ can be decomposed into multiple single-carrier scheduling problem $\mathbf{P_3}$. $\mathbf{P_3}$ consists only of linear objective function and linear constraints with variables $n_{k,j,i}^{f}(t)$ and $n_{k,j,i}^{c,f}(t)$. Thus it is a convex optimization problem.

5.3.1   KKT Conditions for Optimality

For convex optimization problems, the KKT conditions are necessary and sufficient for optimality. Optimal solution for the convex optimization problem $\mathbf{P_3}$ can thus be solved from the KKT conditions given as follows.

$$\sum_{k=1}^{N_u} x_{k,0,i} n_{k,0,i}^{f}(t) + \sum_{j=1}^{N_r}\sum_{k=1}^{N_u} x_{k,j,i} x_{k,j,i}^{c,f}(t) n_{k,j,i}^{c,f}(t) + \sum_{j=1}^{N_r}\sum_{k=1}^{N_u} n_{k,j,i}^{b,f}(t) \le 1$$

$$\text{for } i = 1, \cdots, N_c,\, f = 1, \cdots, F \tag{5.19}$$

$$\sum_{k=1}^{N_u} x_{k,j,i}(1 - x_{k,j,i}^{c,f}(t)) n_{k,j,i}^{f}(t) + \sum_{k=1}^{N_u} x_{k,j,i} x_{k,j,i}^{c,f}(t) n_{k,j,i}^{c,f}(t) + \sum_{k=1}^{N_u} x_{k,j,i} n_{k,j,i}^{b,f}(t) \le 1$$

$$\text{for } i = 1, \cdots, N_c,\, j = 1, \cdots, N_r,\, f = 1, \cdots, F \tag{5.20}$$

$$-n_{k,j,i}^{f}(t) \le 0 \quad \forall k, j, i, f \tag{5.21}$$

$$-n_{k,j,i}^{c,f}(t) \le 0 \quad \forall k, j, i, f \tag{5.22}$$

$$\lambda_i^{f}(t) \ge 0 \quad \forall i, f \tag{5.23}$$

$$\mu_{j,i}^{f}(t) \ge 0 \quad \forall j, f \tag{5.24}$$

$$v_{k,j,i}^{f}(t) \ge 0 \quad \forall k, j, i, f \tag{5.25}$$

$$v_{k,j,i}^{c,f}(t) \ge 0 \quad \forall k, j, i, f \tag{5.26}$$

$$\lambda_i^{f}(t)\Big(\sum_{k=1}^{N_u} x_{k,0,i} n_{k,0,i}^{f}(t) + \sum_{j=1}^{N_r}\sum_{k=1}^{N_u} x_{k,j,i} x_{k,j,i}^{c,f}(t) n_{k,j,i}^{c,f}(t) + \sum_{j=1}^{N_r}\sum_{k=1}^{N_u} x_{k,j,i} n_{k,j,i}^{b,f}(t) - 1\Big) = 0$$

$$\text{for } i = 1, \cdots, N_c,\, f = 1, \cdots, F \tag{5.27}$$

$$\mu_{j,i}^{f}(t)\Big(\sum_{k=1}^{N_u} x_{k,j,i}(1 - x_{k,j,i}^{c,f}(t)) n_{k,j,i}^{f}(t) + \sum_{k=1}^{N_u} x_{k,j,i} x_{k,j,i}^{c,f}(t) n_{k,j,i}^{c,f}(t) + \sum_{k=1}^{N_u} x_{k,j,i} n_{k,j,i}^{b,f}(t) - 1\Big) = 0$$

$$\text{for } i = 1, \cdots, N_c,\, j = 1, \cdots, N_r,\, f = 1, \cdots, F \tag{5.28}$$

$$v_{k,j,i}^{f}(t) n_{k,j,i}^{f}(t) = 0 \quad \forall k, j, i, f \tag{5.29}$$

$$v_{k,j,i}^{c,f}(t) n_{k,j,i}^{c,f}(t) = 0 \quad \forall k, j, i, f \tag{5.30}$$

$$\nabla_{n_{k,0,i}^{f}(t)} L(n_{k,j,i}^{f}(t), n_{k,j,i}^{c,f}(t), \lambda_i^{f}(t), \mu_{j,i}^{f}(t), v_{k,j,i}^{f}(t), v_{k,j,i}^{c,f}(t)) = 0 \quad \forall k, i, f \tag{5.31}$$

$$\nabla_{n_{k,j,i}^{f}(t)} L(n_{k,j,i}^{f}(t), n_{k,j,i}^{c,f}(t), \lambda_i^{f}(t), \mu_{j,i}^{f}(t), v_{k,j,i}^{f}(t), v_{k,j,i}^{c,f}(t)) = 0 \quad \forall k, j, i, f \tag{5.32}$$

$$\nabla_{n_{k,j,i}^{c,f}(t)} L(n_{k,j,i}^{f}(t), n_{k,j,i}^{c,f}(t), \lambda_i^{f}(t), \mu_{j,i}^{f}(t), v_{k,j,i}^{f}(t), v_{k,j,i}^{c,f}(t)) = 0 \quad \forall k, j, i, f, \tag{5.33}$$

where $\lambda_i^{f}(t)$, $\mu_{j,i}^{f}(t)$, and $v_{k,j,i}^{f}(t)$ are the Lagrangian multipliers, and $L(n_{k,j,i}^{f}(t), n_{k,j,i}^{c,f}(t), \lambda_i^{f}(t), \mu_{j,i}^{f}(t), v_{k,j,i}^{f}(t))$ is the Lagrangian of $\mathbf{P}_3$, given as

$$L(n_{k,j,i}^{f}(t), n_{k,j,i}^{c,f}(t), \lambda_i^{f}(t), \mu_{j,i}^{f}(t), v_{k,j,i}^{f}(t), v_{k,j,i}^{c,f}(t))$$

$$= -\sum_{k=1}^{N_u}\sum_{i=1}^{N_c}\sum_{j=0}^{N_r} \frac{1}{R_k(t-1)}\Big(x_{k,j,i}(1 - x_{k,j,i}^{c,f}(t)) R_{k,j,i}^{f}(t) n_{k,j,i}^{f}(t) + x_{k,j,i} x_{k,j,i}^{c,f}(t) R_{k,j,i}^{c,f}(t) n_{k,j,i}^{c,f}(t)\Big)$$

$$+ \sum_{i=1}^{N_c} \lambda_i^f(t) \Big(\sum_{k=1}^{N_u} x_{k,0,i} n_{k,0,i}^f(t) + \sum_{j=1}^{N_r} \sum_{k=1}^{N_u} x_{k,j,i} x_{k,j,i}^{c,f}(t) n_{k,j,i}^{c,f}(t) + \sum_{j=1}^{N_r} \sum_{k=1}^{N_u} x_{k,j,i} n_{k,j,i}^{b,f}(t) - 1 \Big)$$

$$+ \sum_{i=1}^{N_c} \sum_{j=1}^{N_r} \mu_{j,i}^f(t) \Big(\sum_{k=1}^{N_u} x_{k,j,i} (1 - x_{k,j,i}^{c,f}(t)) n_{k,j,i}^f(t)$$

$$+ \sum_{k=1}^{N_u} x_{k,j,i} x_{k,j,i}^{c,f}(t) n_{k,j,i}^{c,f}(t) + \sum_{k=1}^{N_u} x_{k,j,i} n_{k,j,i}^{b,f}(t) - 1 \Big)$$

$$- \sum_{i=1}^{N_c} \sum_{j=0}^{N_r} \sum_{k=1}^{N_u} v_{k,j,i}^f(t) n_{k,j,i}^f(t) - \sum_{i=1}^{N_c} \sum_{j=1}^{N_r} \sum_{k=1}^{N_u} v_{k,j,i}^{c,f}(t) n_{k,j,i}^{c,f}(t). \tag{5.34}$$

From (5.31) to (5.33), we have

$$-\frac{1}{R_k(t-1)} x_{k,0,i} R_{k,0,i}^f(t) + \lambda_i^f(t) x_{k,0,i} - v_{k,0,i}^f(t) = 0, \tag{5.35}$$

$$x_{k,j,i} \big(1 - x_{k,j,i}^{c,f}(t)\big) \Big(-\frac{1}{R_k(t-1)} R_{k,j,i}^f(t) + \lambda_i^f(t) a_{k,j,i}^f(t) + \mu_{j,i}^f(t) \big(1 + a_{k,j,i}^f(t)\big) \Big) - v_{k,j,i}^f(t) = 0, \tag{5.36}$$

and

$$x_{k,j,i} x_{k,j,i}^{c,f}(t) \Big(-\frac{1}{R_k(t-1)} R_{k,j,i}^{c,f}(t) + \lambda_i^f(t) \big(1 + a_{k,j,i}^{c,f}(t)\big) + \mu_{j,i}^f(t) \big(1 + a_{k,j,i}^{c,f}(t)\big) \Big) - v_{k,j,i}^{c,f}(t) = 0. \tag{5.37}$$

From the KKT conditions (5.19)–(5.30) and (5.35)–(5.37), the optimization problem can be decoupled into N_c independent optimization problems, one for each sector. Therefore, the resource allocation problem can be solved independently for each sector. In the following section, without loss of generality, the resource allocation problem can be solved analytically for the ith sector. The resultant resource allocation strategy is applicable to all the other sectors in the network.

5.3.2   Optimal Resource Allocation Strategy Based on Solving the KKT Conditions

The goal is to find the optimal M-UE index k_0^*, R-UE index $k_{1,j}^*$, and C-UE index k_2^* to be served by the MeNB of the ith sector, the jth RN in the ith sector, and their cooperation in each RB of a subframe, and the corresponding optimal $n_{k,j,i}^{f*}(t)$ and $n_{k,j,i}^{c,f*}(t)$ values. With the optimal $n_{k,j,i}^{f*}(t)$ and $n_{k,j,i}^{c,f*}(t)$ values from all RBs in all the sectors, the optimal T_b value for scheduling direct/access link transmission and backhaul transmission can be decided. Towards this end, the KKT conditions given in (5.19)–(5.30) and (5.35)–(5.37) can first be solved, and the optimal Lagrangian multiplier values can be obtained as follows.

$$\lambda_i^{f*}(t) = \max\{\lambda_{i,A}^f(t), \lambda_{i,B}^f(t), \lambda_{i,C}^f(t)\}, \tag{5.38}$$

$$\mu_{j,i}^{f*}(t) = \left(\tilde{\mu}_{j,i}^f(t)\right)^+, \tag{5.39}$$

$$v_{k_0,j,i}^{f*}(t) = \lambda_i^{f*}(t) - \frac{R_{k_0,0,i}^f(t)}{R_{k_0}(t-1)} \quad \text{for } k_0 \in \mathcal{K}_{0,i}, \tag{5.40}$$

$$v_{k_{1,j},j,i}^{f*}(t) = (1 + a_{k_{1,j},j,i}^f(t))\left(\mu_{j,i}^{f*}(t) - \frac{1}{1 + a_{k_{1,j},j,i}^f(t)}\left(\frac{R_{k_{1,j},j,i}^f(t)}{R_{k_{1,j}}(t-1)} - a_{k_{1,j},j,i}^f(t)\lambda_i^{f*}(t)\right)\right)$$
$$\text{for } k_{1,j} \in \mathcal{K}_{j,i}, \tag{5.41}$$

and

$$v_{k_2,j,i}^{c,f*}(t) = \frac{1 + a_{k_2,j,i}^{c,f}(t)}{1 + a_{k_{1,j}^*,j,i}^f(t)}\left(\lambda_i^f(t) - \left(\left(\frac{R_{k_2,j,i}^{c,f}(t)(1 + a_{k_{1,j}^*,j,i}^f(t))}{R_{k_2}(t-1)(1 + a_{k_2,j,i}^{c,f}(t))} - \frac{R_{k_{1,j}^*,j,i}^f(t)}{R_{k_{1,j}^*}(t-1)}\right) - v_{k_{1,j}^*,j,i}^f(t)\right)\right)$$
$$\text{for } k_2 \in \mathcal{K}_{j,i}^c, \tag{5.42}$$

where

$$\lambda_{i,A}^f(t) = \max_{k_0 \in \mathcal{K}_{0,i}} \frac{R_{k_0,0,i}^f(t)}{R_{k_0}(t-1)}, \tag{5.43}$$

$$\lambda_{i,B}^f(t) = \max_{j \in \mathcal{J}_{i,1}}\left(\max_{k_2 \in \mathcal{K}_{j,i}^c} \frac{R_{k_2,j,i}^{c,f}(t)(1 + a_{k_{1,j}^*,j,i}^f(t))}{R_{k_2}(t-1)(1 + a_{k_2,j,i}^{c,f}(t))} - \frac{R_{k_{1,j}^*,j,i}^f(t)}{R_{k_{1,j}^*}(t-1)}\right), \tag{5.44}$$

$$\lambda_{i,C}^f(t) = \max_{j \in \mathcal{J}_{i,2}}\max_{k_2 \in \mathcal{K}_{j,i}^c} \frac{R_{k_2,j,i}^{c,f}(t)}{R_{k_2}(t-1)(1 + a_{k_2,j,i}^{c,f}(t))}, \tag{5.45}$$

$$k_{1,j}^* = \arg\max_{k_{1,j} \in \mathcal{K}_{j,i}} \frac{1}{1 + a_{k_{1,j},j,i}^f(t)}\left(\frac{R_{k_{1,j},j,i}^f(t)}{R_{k_{1,j}}(t-1)} - a_{k_{1,j},j,i}^f(t)\lambda_i^f(t)\right), \tag{5.46}$$

$$\tilde{\mu}_{j,i}^f(t) = \max_{k_{1,j} \in \mathcal{K}_{j,i}} \frac{1}{1 + a_{k_{1,j},j,i}^f(t)}\left(\frac{R_{k_{1,j},j,i}^f(t)}{R_{k_{1,j}}(t-1)} - a_{k_{1,j},j,i}^f(t)\lambda_i^f(t)\right), \tag{5.47}$$

and $\mathscr{J}_{i,1} = \{j \,|\, j \in \mathscr{J}_i, \tilde{\mu}_j^f(t) \geq 0\}$, $\mathscr{J}_{i,2} = \{j \,|\, j \in \mathscr{J}_i, \tilde{\mu}_j^f(t) < 0\}$. Here, $\mathscr{J}_{i,1}$ is a set of RNs whose proportional fairness gain from supporting a R-UE is higher than the proportional fairness loss due to backhaul resource consumption at its donor MeNB. $\mathscr{J}_{i,2}$ is a set of RNs whose proportional fairness gain from supporting a R-UE is not high enough to compensate for the proportional fairness loss due to backhaul resource consumption at its donor MeNB. Therefore, it is beneficial for the RNs in $\mathscr{J}_{i,1}$ to transmit to their R-UEs over the fth RB while allowing the RNs in $\mathscr{J}_{i,2}$ to transmit to their R-UEs would lead to loss in overall utility of the network.

In the following context, detailed steps in deriving the optimal Lagrangian multipliers will be shown, followed by more insights on the meaning of $\mathscr{J}_{i,1}$ and $\mathscr{J}_{i,2}$.

From (5.35), it has

$$\lambda_i^f(t) = \frac{R_{k_0,0,i}^f(t)}{R_{k_0}(t-1)} + v_{k_0,0,i}^f(t) \text{ for } k_0 \in \mathscr{K}_{0,i}. \tag{5.48}$$

From (5.36) and (5.37), then

$$\lambda_i^f(t) = \frac{1}{1+a_{k_2,j,i}^{c,f}(t)}\left(\frac{R_{k_2,j,i}^{c,f}(t)(1+a_{k_1,j,j,i}^f(t))}{R_{k_2}(t-1)} - \frac{R_{k_1,j,i}^f(t)(1+a_{k_2,j,i}^{c,f}(t))}{R_{k_1,j}(t-1)}\right)$$

$$+\frac{1}{1+a_{k_2,j,i}^{c,f}(t)}\left(v_{k_2,j,i}^{c,f}(t)\left(1+a_{k_1,j,j,i}^f(t)\right) - v_{k_1,j,j,i}^f(t)\left(1+a_{k_2,j,i}^{c,f}(t)\right)\right)$$

$$\text{for } k_{1,j} \in \mathscr{K}_{j,i}, k_2 \in \mathscr{K}_{j,i}^c, j \in \mathscr{J}_i. \tag{5.49}$$

For all the $k_0 \in \mathscr{K}_{0,i}$, $k_{1,j} \in \mathscr{K}_{j,i}$, $k_2 \in \mathscr{K}_{j,i}^c$ and $j \in \mathscr{J}_i$, the values of $v_{k_0,0,i}^f(t)$, $v_{k_1,j,j,i}^f(t)$ and $v_{k_2,j,i}^f(t)$ should be chosen to ensure a same $\lambda_i^f(t)$ value under different $R_{k_0,j,i}^f(t)$, $R_{k,j,i}^f(t)$ and $R_{k,j,i}^{c,f}(t)$ values. Since from (5.48) and (5.49), then

$$v_{k_0,j,i}^f(t) = \lambda_i^f(t) - \frac{R_{k_0,0,i}^f(t)}{R_{k_0}(t-1)}, \tag{5.50}$$

and

$$v_{k_2,j,i}^{c,f}(t) = \frac{1+a_{k_2,j,i}^{c,f}(t)}{1+a_{k_1,j,j,i}^f(t)}\left(\lambda_i^f(t) - \left(\left(\frac{R_{k_2,j,i}^{c,f}(t)(1+a_{k_1,j,j,i}^f(t))}{R_{k_2}(t-1)(1+a_{k_2,j,i}^{c,f}(t))} - \frac{R_{k_1,j,i}^f(t)}{R_{k_1,j}(t-1)}\right)\right.\right.$$

$$\left.\left. -v_{k_1,j,j,i}^f(t)\right)\right), \tag{5.51}$$

and the KKT conditions in (5.25) and (5.26) require $v_{k_0,j,i}^f(t) \geq 0$ and $v_{k_2,j,i}^{c,f}(t) \geq 0$, the $\lambda_i^f(t)$ value should be chosen as

$$
\lambda_i^f(t) = \max_{k_0 \in \mathcal{K}_{0,i}, k_1 \in \mathcal{K}_{j,i}, k_2 \in \mathcal{K}_{j,i}^c, j \in \mathcal{J}_i} \left\{ \frac{R_{k_0,0,i}^f(t)}{R_{k_0}(t-1)} , \right.
$$
$$
\left. \left(\frac{R_{k_2,j,i}^{c,f}(t)(1 + a_{k_1,j,i}^f(t))}{R_{k_2}(t-1)(1 + a_{k_2,j,i}^{c,f}(t))} - \frac{R_{k_1,j,i}^f(t)}{R_{k_1}(t-1)} \right) - v_{k_1,j,i}^f(t) \right\}, \quad (5.52)
$$

By such choice of $\lambda_i^f(t)$ value, the k_0^*th B-UE or the k_2^*th C-UE that achieves the $\lambda_i^f(t)$ value could have $v_{k_0^*,j,i}^f(t) = 0$ or $v_{k_2^*,j,i}^{c,f}(t) = 0$. By the KKT conditions (5.29) and (5.30), it could have $n_{k_0^*,j,i}^f(t) > 0$ or $n_{k_2^*,j,i}^{f,c}(t) > 0$, i.e., RB f is allocated to the UE.

From (5.52), we see that $\lambda_i^f(t)$ depends on the value of $v_{k_1,j,i}^f(t)$ which can be derived from (5.36) as

$$
v_{k_1,j,i}^f(t) = (1 + a_{k_1,j,i}^f(t)) \left(\mu_{j,i}^f(t) - \frac{1}{1 + a_{k_1,j,i}^f(t)} \left(\frac{R_{k_1,j,i}^f(t)}{R_{k_1}(t-1)} - a_{k_1,j,i}^f(t)\lambda_i^f(t) \right) \right)
$$
$$
\text{for } k_{1,j} \in \mathcal{K}_{j,i}. \quad (5.53)
$$

Since the KKT condition (5.25) requires $v_{k_1,j,i}^f(t) \geq 0$, the value of $\mu_{j,i}^f(t)$ should be chosen as

$$
\mu_{j,i}^f(t) = \max_{k_{1,j} \in \mathcal{K}_{j,i}} \frac{1}{1 + a_{k_1,j,i}^f(t)} \left(\frac{R_{k_1,j,i}^f(t)}{R_{k_1}(t-1)} - a_{k_1,j,i}^f(t)\lambda_i^f(t) \right) + \Delta_{j,i}^f(t)
$$
$$
= \max_{k_{1,j} \in \mathcal{K}_{j,i}} \tilde{\mu}_{j,i}^f(t) + \Delta_{j,i}^f(t), \quad (5.54)
$$

where $\Delta_{j,i}^f(t) \geq 0$ is a parameter chosen to ensure $\mu_{j,i}^f(t) \geq 0$. For the case with $\tilde{\mu}_{j,i}^f(t) \geq 0$, we set $\Delta_{j,i}^f(t) = 0$ such that we could have $v_{k_{1,j},j,i}^f(t) = 0$ and $n_{k_1^*,j,i}^f(t) > 0$ for the R-UE with index

$$
k_{1,j}^* = \arg \max_{k_{1,j} \in \mathcal{K}_{j,i}} \frac{1}{1 + a_{k_1,j,i}^f(t)} \left(\frac{R_{k_1,j,i}^f(t)}{R_{k_1}(t-1)} - a_{k_1,j,i}^f(t)\lambda_i^f(t) \right), \quad (5.55)
$$

i.e., the fth RB could be assigned to the $k_{1,j}^*$th R-UE. For the case with $\tilde{\mu}_{j,i}^f(t) < 0$, the value of $\Delta_{j,i}^f(t)$ should be chosen as

$$\Delta_{j,i}^{f}(t) = \frac{1}{1 + a_{k_{1,j}^{*},j,i}^{f}(t)} \left(a_{k_{1,j}^{*},j,i}^{f}(t)\lambda_{i}^{f}(t) - \frac{R_{k_{1,j}^{*},j,i}^{f}(t)}{R_{k_{1,j}^{*}}(t-1)} \right) \qquad (5.56)$$

to ensure $\mu_{j,i}^{t}(t) \geq 0$.

Substituting $v_{k_{1,j},j,i}^{f}(t)$ obtained under the two instances of $\mu_{j,i}^{f}(t)$ into (5.52), then the optimal $\lambda_{i}^{f*}(t)$ expression given in (5.38).

It can be seen that $\lambda_{i,A}^{f}(t)$, $\lambda_{i,B}^{f}(t)$ and $\lambda_{i,C}^{f}(t)$ represent the gains in proportional fairness value at the ith MeNB by different strategies in assigning the fth RB at time t. Specifically, $\lambda_{i,A}^{f}(t)$ calculates the gain in assigning RB f to the best M-UE. $\lambda_{i,B}^{f}(t)$ calculates the gain in assigning RB f to the best C-UE in the case where the C-UE locates in the coverage range of the RN with index $j \in \mathcal{J}_{i,1}$. $\lambda_{i,C}^{f}(t)$ calculates the gain in assigning RB f to the best C-UE in the case where the C-UE locates in the coverage range of the RN with index $j \in \mathcal{J}_{i,2}$. The value of $\lambda_{i}^{f}(t)$ is chosen to be the highest gain among all the gains under the different strategies, and the corresponding winning UE is assigned with the RB.

Based on the obtained $\lambda_{i}^{f}(t)$ value, the value of $\mu_{j,i}^{f}(t)$ can be calculated from (5.39). It can be seen that $\tilde{\mu}_{j,i}^{f}(t)$ represents the gain in proportional fairness value at the jth RN in the ith sector by serving the R-UEs. Specifically, the term $R_{k_{1,j},j,i}^{f}(t)/(R_{k_{1,j}}(t-1)(1 + a_{k_{1,j},j,i}^{f}(t)))$ is the gain of serving the R-UE $k_{1,j}$ by the jth RN and the term $a_{k_{1,j},j,i}^{f}(t)\lambda_{i}^{f}(t)/(1 + a_{k_{1,j},j,i}^{f}(t))$ is the cost of using the backhaul link from the ith MeNB to the jth RN in serving R-UE $k_{1,j}$ R-UE, i.e., if the amount of $a_{k_{1,j},j,i}^{f}(t)/(1 + a_{k_{1,j},j,i}^{f}(t))$ radio resource used on the backhaul link is instead used by the ith MeNB to transmit to its M-UE or C-UE, an amount of $a_{k_{1,j},j,i}^{f}(t)\lambda_{i}^{f}(t)/(1 + a_{k_{1,j},j,i}^{f}(t))$ gain can be achieved. As such, $\tilde{\mu}_{j,i}^{f}(t) \geq 0$ indicates that the gain in serving the R-UEs is no less than the cost in using the backhaul link. In this case, it is beneficial to serve the R-UE. $\tilde{\mu}_{j,i}^{f}(t) < 0$ indicates that the gain in serving the R-UE is less than the cost in using the backhaul link. RN receives no gain in serving the R-UE, i.e., $\mu_{j,i}^{f}(t) = 0$. It is better not to serve the R-UE.

Based on the derived optimal Lagrangian multiplier values, consider the following two cases in finding the optimal k_{0}^{*}, $k_{1,j}^{*}$, k_{2}^{*}, $n_{k,j,i}^{f*}(t)$, and $n_{k,j,i}^{c,f*}(t)$ values.

Case 1. $\lambda_{i,A}^{f}(t) \geq \max\{\lambda_{i,B}^{f}(t), \lambda_{i,C}^{f}(t)\}$

In this case, it has

$$\lambda_{i}^{f}(t) = \max_{k_{0} \in \mathcal{K}_{0,i}} \frac{R_{k_{0},0,i}^{f}(t)}{R_{k_{0}}(t-1)}, \qquad (5.57)$$

and

$$
\max_{k_2 \in \mathscr{H}^c_{j_1^*,i}} \frac{R^{c,f}_{k_2,j_1^*,i}(t)}{R_{k_2}(t-1)(1+a^{c,f}_{k_2,j_1^*,i}(t))} < \max_{k_0 \in \mathscr{K}_{0,i}} \frac{R^f_{k_0,0,i}(t)}{R_{k_0}(t-1)(1+a^f_{k_{1,j},j_1^*,i}(t))}
$$

$$
+ \frac{R^f_{k_{1,j}^*,j_1^*,i}(t)}{R_{k_{1,j}^*}(t-1)(1+a^f_{k_{1,j}^*,j_1^*,i}(t))}, \quad (5.58)
$$

$$
\max_{k_2 \in \mathscr{H}^c_{j_2^*,i}} \frac{R^{c,f}_{k_2,j_2^*,i}(t)}{R_{k_2}(t-1)(1+a^{c,f}_{k_2,j_2^*,i}(t))} < \max_{k_0 \in \mathscr{K}_{0,i}} \frac{R^f_{k_0,0,i}(t)}{R_{k_0}(t-1)}, \quad (5.59)
$$

where

$$
j_1^* = \arg \max_{j \in \mathscr{J}_{i,1}} \left(\max_{k_2 \in \mathscr{H}^c_{j,i}} \frac{R^{c,f}_{k_2,j,i}(t)(1+a^f_{k_{1,j}^*,j,i}(t))}{R_{k_2}(t-1)(1+a^{c,f}_{k_2,j,i}(t))} - \frac{R^f_{k_{1,j}^*,j,i}(t)}{R_{k_{1,j}^*}(t-1)} \right), \quad (5.60)
$$

and

$$
j_2^* = \arg \max_{j \in \mathscr{J}_{i,2}} \max_{k_2 \in \mathscr{H}^c_{j,i}} \frac{R^{c,f}_{k_2,j,i}(t)}{R_{k_2}(t-1)(1+a^{c,f}_{k_2,j,i}(t))}, \quad (5.61)
$$

which correspond to the indices of the RNs leading to $\lambda_{i,B}$ and $\lambda_{i,C}$, respectively. The left side of the inequalities (5.58) and (5.59) is the proportional fairness gain by serving the C-UEs in the fth RB. The right side of (5.58) is the proportional fairness gain by serving the M-UEs and the R-UEs associated with the RNs in $\mathscr{J}_{i,1}$. The right side of (5.59) is the proportional fairness gain by serving the M-UEs. From (5.58) and (5.59), the case with $\lambda^f_{i,A}(t) \geq \max\{\lambda^f_{i,B}(t), \lambda^f_{i,C}(t)\}$ corresponds to the scenario where serving the C-UE cooperatively on RB f by the MeNB and the RN receives a less gain than using the RB for the respective M-UE and the R-UE separately. In another word, CoMP is not used on RB f.

Substituting (5.57) into (5.39)–(5.42), then $v^{c,f*}_{k_2,j,i}(t) > 0$ for all $k_2 \in \mathscr{H}^c_{j,i}$,

$$
v^{f*}_{k_0,j,i}(t) = \begin{cases} 0 & \text{if } k_0 = k_0^* \\ \dfrac{R^f_{k_0^*,0,i}(t)}{R_{k_0^*}(t-1)} - \dfrac{R^f_{k_0,0,i}(t)}{R_{k_0}(t-1)} & \text{if } k_0 \in \mathscr{K}_{0,i}, k_0 \neq k_0^* \end{cases} \quad (5.62)
$$

where

$$
k_0^* = \arg \max_{k_0 \in \mathscr{K}_{0,i}} \frac{R^f_{k_0,0,i}(t)}{R_{k_0}(t-1)}, \quad (5.63)
$$

$$\mu_{j,i}^{f*}(t) = \begin{cases} \dfrac{1}{1+a_{k_{1,j}^*,j,i}^f(t)}\left(\dfrac{R_{k_{1,j}^*,j,i}^f(t)}{R_{k_{1,j}^*}(t-1)} - a_{k_{1,j}^*,j,i}^f(t)\lambda_i^f(t)\right) & \text{if } j \in \mathscr{J}_{i,1}, \\[4mm] 0 & \text{if } j \in \mathscr{J}_{i,2}, \end{cases} \quad (5.64)$$

where

$$k_{1,j}^* = \arg\max_{k_{1,j}\in\mathscr{K}_{j,i}} \frac{1}{1+a_{k_{1,j},j,i}^f(t)}\left(\frac{R_{k_{1,j},j,i}^f(t)}{R_{k_{1,j}}(t-1)} - a_{k_{1,j},j,i}^f(t)\lambda_i^f(t)\right), \quad (5.65)$$

and

$$v_{k_{1,j},j,i}^{f*}(t)$$
$$= \begin{cases} 0 & \text{if } k_{1,j}=k_{1,j}^*, j \in \mathscr{J}_{i,1} \\[3mm] (1+a_{k_{1,j},j,i}^f(t))\left(\mu_j^{f*}(t) - \dfrac{1}{1+a_{k_{1,j},j,i}^f(t)}\left(\dfrac{R_{k_{1,j},j,i}^f(t)}{R_{k_{1,j}}(t-1)} - a_{k_{1,j},j,i}^f(t)\lambda_i^f(t)\right)\right) & \text{otherwise.} \end{cases}$$
$$(5.66)$$

According to the KKT conditions (5.29) and (5.30), $n_{k_0,j,i}^f(t) > 0$, $n_{k_{1,j},j,i}^f(t) > 0$ or $n_{k_2,j,i}^{c,f}(t) > 0$ only when $v_{k_0,j,i}^f(t) = 0$, $v_{k_{1,j},j,i}^f(t) = 0$ or $v_{k_2,j,i}^{c,f}(t) = 0$. Therefore, in the case with $\lambda_{i,A}^f(t) \geq \max\{\lambda_{i,B}^f(t), \lambda_{i,C}^f(t)\}$, the optimal strategy in allocating the fth RB at the tth subframe in the ith MeNB is to let the MeNB transmit to the k_0^*th M-UE on the entire RB f, the j_1th ($j_1 \in \mathscr{J}_{i,1}$) RN transmit to the $k_{1,j}^*$th R-UE on the entire RB f, and the j_2th ($j_2 \in \mathscr{J}_{i,2}$) RN not serve any of its R-UEs.

The optimal $n_{k,j,i}^f(t)$ and $n_{k,j,i}^{c,f}(t)$ values for the virtual resource allocation problem can be solved from the constraints (5.10) and (5.11) as

$$n_{k,j,i}^{c,f*} = 0, \; n_{k,0,i}^{f*} = 0 \text{ for } k \neq k_0^*, \; n_{k,j,i}^{f*} = 0 \text{ for } k \neq k_{1,j}^* \quad (5.67)$$

$$n_{k_{j,1}^*,j,i}^{f*}(t) = \begin{cases} \dfrac{1}{1+a_{k_{j,1}^*,j,i}^f(t)} & \text{if } j \in \mathscr{J}_{i,1} \\[3mm] 0 & \text{if } j \in \mathscr{J}_{i,2}, \end{cases} \quad (5.68)$$

and

$$n_{k_0^*,0,i}^{f*}(t) = \left(1 - \sum_{j\in\mathscr{J}_{i,1}} \frac{a_{k_{j,1}^*,j,i}^f(t)}{1+a_{k_{j,1}^*,j,i}^f(t)}\right)^+. \quad (5.69)$$

Case 2. $\lambda_{i,A}^{f}(t) < \max\{\lambda_{i,B}^{f}(t), \lambda_{i,C}^{f}(t)\}$

In this case, it has

$$
\lambda_i^{f*}(t) = \max\left\{ \max_{j \in \mathscr{J}_{i,1}} \left(\max_{k_2 \in \mathscr{K}_{j,i}^c} \frac{R_{k_2,j,i}^{c,f}(t)(1 + a_{k_1^*,j,i}^f(t))}{R_{k_2}(t-1)(1 + a_{k_2,j,i}^{c,f}(t))} - \frac{R_{k_1^*,j,i}^f(t)}{R_{k_1^*}(t-1)} \right), \right.
$$
$$
\left. \max_{j \in \mathscr{J}_{i,2}} \left(\max_{k_2 \in \mathscr{K}_{j,i}^c} \frac{R_{k_2,j,i}^{c,f}(t)}{R_{k_2}(t-1)(1 + a_{k_2,j,i}^{c,f}(t))} \right) \right\}. \tag{5.70}
$$

Following the same analysis as in Case 1, that the case with $\lambda_{i,A}^{f}(t) < \max\{\lambda_{i,B}^{f}(t), \lambda_{i,C}^{f}(t)\}$ corresponds to a scenario where the gain in proportional fairness value by serving the C-UEs cooperatively on RB f by the MeNB and the RN is higher than the gain in using RB f to serve the M-UEs and the R-UEs separately.

Substituting (5.70) into (5.39)–(5.42), then $v_{k_0,j,i}^f(t) > 0$ for all $k_0 \in \mathscr{K}_{0,i}$, $\mu_{j,i}^{f*}(t)$, $k_{1,j}^*$, and $v_{k_1,j,j,i}^{f*}(t)$ values as given in (5.64)–(5.66), and

$$
v_{k_2,j,i}^{c,f*}(t) = \begin{cases} 0 & \text{if } k_2 = k_2^* \\ \dfrac{1 + a_{k_2,j,i}^{c,f}(t)}{1 + a_{k_{1,j}^*,j,i}^f(t)} \left(\lambda_i^f(t) - \left(\dfrac{R_{k_2,j,i}^{c,f}(t)(1 + a_{k_{1,j}^*,j,i}^f(t))}{R_{k_2}(t-1)(1 + a_{k_2,j,i}^{c,f}(t))} - \dfrac{R_{k_{1,j}^*,j,i}^f(t)}{R_{k_{1,j}^*}(t-1)} \right) \right) \\ \qquad \text{if } k_2 \in \mathscr{K}_{j,i}, k_2 \neq k_2^*, j \in \mathscr{J}_{i,1} \\ (1 + a_{k_2,j,i}^{c,f}(t)) \left(\lambda_i^f(t) - \dfrac{R_{k_2,j,i}^{c,f}(t)}{R_{k_2}(t-1)(1 + a_{k_2,j,i}^{c,f}(t))} \right) \text{ if } k_2 \in \mathscr{K}_{j,i}, k_2 \neq k_2^*, j \in \mathscr{J}_{i,2}, \end{cases}
$$
$$\tag{5.71}$$

where

$$
k_2^* = \begin{cases} \arg\max_{k_2 \in \mathscr{K}_{j^*,i}^c} \dfrac{R_{k_2,j^*,i}^{c,f}(t)(1 + a_{k_{1,j}^*,j^*,i}^f(t))}{R_{k_2}(t-1)(1 + a_{k_2,j^*,i}^{c,f}(t))} & \text{if } \lambda_{i,B}^f(t) \geq \lambda_{i,C}^f(t) \\ \arg\max_{k_2 \in \mathscr{K}_{j^*,i}^c} \dfrac{R_{k_2,j^*,i}^{c,f}(t)}{R_{k_2}(t-1)(1 + a_{k_2,j^*,i}^{c,f}(t))} & \text{if } \lambda_{i,B}^f(t) < \lambda_{i,C}^f(t), \end{cases} \tag{5.72}
$$

and

$$
j^* = \begin{cases} \arg\max_{j \in \mathscr{J}_{i,1}} \left(\max_{k_2 \in \mathscr{K}_{j,i}^c} \dfrac{R_{k_2,j,i}^{c,f}(t)(1 + a_{k_{1,j}^*,j,i}^f(t))}{R_{k_2}(t-1)(1 + a_{k_2,j,i}^{c,f}(t))} - \dfrac{R_{k_{1,j}^*,j,i}^f(t)}{R_{k_{1,j}^*}(t-1)} \right) & \text{if } \lambda_{i,B}^f(t) \geq \lambda_{i,C}^f(t) \\ \arg\max_{j \in \mathscr{J}_{i,2}} \left(\max_{k_2 \in \mathscr{K}_{j,i}^c} \dfrac{R_{k_2,j,i}^{c,f}(t)}{R_{k_2}(t-1)(1 + a_{k_2,j,i}^{c,f}(t))} \right) & \text{if } \lambda_{i,B}^f(t) < \lambda_{i,C}^f(t). \end{cases} \tag{5.73}
$$

According to the KKT conditions (5.29) and (5.30), in order to have $n_{k_0,j,i}^f(t) >$
$0, n_{k_1,j,i}^f t > 0$ or $n_{k_2,j,i}^{c,f}(t) > 0, v_{k_0,j,i}^f(t) = 0, v_{k_1,j,i}^{c,f}(t) = 0$ or $v_{k_2,j,i}^{c,f}(t) = 0$ is
required. Based on the above derivations, we have that $n_{k_2^*,j^*,i}^{c,f}(t) > 0, n_{k_1^*,j,i}^f(t) >$
0, and all the other $n_{k,j,i}^f(t)$ values are zero. The optimal resource allocation strategy
at the tth subframe for the case with $\lambda_{i,A}^f(t) < \max\{\lambda_{i,B}^f(t), \lambda_{i,C}^f(t)\}$ is therefore to
allocate the fth RB of the ith MeNB and the j^*th RN to jointly serve the k_2^*th
C-UE, allocate the fth RB of the RN with index $j \in \mathcal{J}_{i,1}, j \neq j^*$ to serve the
$k_{1,j}^*$th R-UE, and let the RN with index $j \in \mathcal{J}_{i,2}$ not serve any of the UEs in the
fth RB.

The optimal $n_{k,j,i}^f(t)$ and $n_{k,j,i}^{c,f}(t)$ values for the virtual resource allocation
problem can be solved from the constraints (5.10) and (5.11) as

$$n_{k,0,i}^{f*} = 0, \ n_{k,j,i}^{c,f*} = 0 \text{ for } k \neq k_2^*, \ n_{k,j,i}^{f*} = 0 \text{ for } k \neq k_{1,j}^* \tag{5.74}$$

$$n_{k_{j,1}^*,j,i}^f(t) = \begin{cases} \dfrac{1}{1+a_{k_{1,j}^*,j,i}^f(t)} & \text{if } j \in \mathcal{J}_{i,1}/\{j^*\} \\ 0 & \text{if } j \in \mathcal{J}_{i,2}, \end{cases} \tag{5.75}$$

and

$$n_{k_2^*,j,i}^{c,f}(t) = \frac{1}{1+a_{k_2^*,j^*,i}^f} \left(1 - \sum_{j \in \mathcal{J}_{i,1}/\{j^*\}} \frac{a_{k_{1,j}^*,j,i}^f(t)}{1+a_{k_{1,j}^*,j,i}^f(t)}\right)^+. \tag{5.76}$$

The above analysis can be summarized in the following proposition on the asymptotically optimal resource allocation based on fairness consideration.

Proposition 5.1. *It is asymptotically optimal to allocate radio resources for het-*
erogeneous relay networks with intra-cell CoMP and proportional fairness consid-
eration using the following strategy. For the fth RB in the tth subframe in the ith
sector,

1. *When $\lambda_{i,A}^f(t) \geq \max\{\lambda_{i,B}^f(t), \lambda_{i,C}^f(t)\}$, it is optimal to let the ith MeNB serve the*
   k_0^*th M-UE over the whole RB, the RNs with indices $j \in \mathcal{J}_{i,1}$ serve the $k_{1,j}^*$th
   R-UE over the whole RB, and the RNs with indices $j \in \mathcal{J}_{i,2}$ in idle where

$$k_0^* = \arg \max_{k_0 \in \mathcal{K}_{0,i}} \frac{R_{k_0,0,i}^f(t)}{R_{k_0}(t-1)}, \tag{5.77}$$

and

$$k_{1,j}^* = \arg \max_{k_{1,j} \in \mathcal{K}_{j,i}} \frac{1}{1+a_{k_{1,j},j,i}^f(t)} \left(\frac{R_{k_{1,j},j,i}^f(t)}{R_{k_{1,j}}(t-1)} - a_{k_{1,j},j,i}^f(t)\lambda_i^f(t)\right), \tag{5.78}$$

2. When $\lambda_{i,A}^{f}(t) < \max\{\lambda_{i,B}^{f}(t), \lambda_{i,C}^{f}(t)\}$, it is optimal to let the ith MeNB and the j^{*}th RN cooperatively serve the k_{2}^{*}th C-UE, the RN with index $j \in \mathcal{J}_{i,1}/\{j^{*}\}$ serve the $k_{1,j}^{*}$th R-UE, and the RN with index $j \in \mathcal{J}_{i,2}/\{j^{*}\}$ in idle, where $k_{1,j}^{*}$ is given in (5.78),

$$
j^{*} = \begin{cases} \arg\max_{j \in \mathcal{J}_{i,1}} \left(\max_{k_2 \in \mathcal{K}_{j,i}^{c}} \dfrac{R_{k_2,j,i}^{c,f}(t)(1+a_{k_{1,j}^{*},j,i}^{f}(t))}{R_{k_2}(t-1)(1+a_{k_2,j,i}^{c,f}(t))} - \dfrac{R_{k_{1,j}^{*},j,i}^{f}(t)}{R_{k_{1,j}^{*}}(t-1)} \right) & \text{if } \lambda_{i,B}^{f}(t) \geq \lambda_{i,C}^{f}(t) \\[4mm] \arg\max_{j \in \mathcal{J}_{i,2}} \left(\max_{k_2 \in \mathcal{K}_{j,i}^{c}} \dfrac{R_{k_2,j,i}^{c,f}(t)}{R_{k_2}(t-1)(1+a_{k_2,j,i}^{c,f}(t))} \right) & \text{if } \lambda_{i,B}^{f}(t) < \lambda_{i,C}^{f}(t), \end{cases}
$$

$$(5.79)$$

and

$$
k_{2}^{*} = \begin{cases} \arg\max_{k_2 \in \mathcal{K}_{j^*,i}^{c}} \dfrac{R_{k_2,j^*,i}^{c,f}(t)(1+a_{k_{1,j}^{*},j^*,i}^{f}(t))}{R_{k_2}(t-1)(1+a_{k_2,j^*,i}^{c,f}(t))} & \text{if } \lambda_{i,B}^{f}(t) \geq \lambda_{i,C}^{f}(t) \\[4mm] \arg\max_{k_2 \in \mathcal{K}_{j^*,i}^{c}} \dfrac{R_{k_2,j^*,i}^{c,f}(t)}{R_{k_2}(t-1)(1+a_{k_2,j^*,i}^{c,f}(t))} & \text{if } \lambda_{i,B}^{f}(t) < \lambda_{i,C}^{f}(t). \end{cases}
$$

$$(5.80)$$

Proposition 5.1 provides a guideline on fairly allocating the network resources. The basic criterion in assigning a RB to a UE is to ensure a maximal proportional fairness profit, where the profit by supporting a UE is calculated as the proportional fairness gain obtained by serving the UE minus the cost. When supporting a R-UE, the data information should be first conveyed from the MeNB to the RN via the backhaul link. The radio resource consumed by the backhaul transmission can be otherwise used by the MeNB to transmit to its M-UEs or help with the C-UEs. Therefore, the cost in supporting the R-UE is the proportional fairness loss at the MeNB due to backhaul resource consumption. If the gain in supporting any of its R-UEs in a particular RB cannot cover the cost, then the RN would not serve any of its R-UEs in that RB. When supporting a C-UE, besides backhaul resource consumption, the C-UE consumes resources from both the MeNB and the RN which could be otherwise used for serving their respective M-UEs and R-UEs. If it counts the gain in supporting the C-UE as the gain of the MeNB, the cost will be the proportional fairness loss at the RN due to deprive of the RB from the R-UEs as well as the backhaul resource consumption. The $\lambda_{i}^{f}(t)$ and the $\mu_{i,j}^{f}(t)$ values used in deriving Proposition 5.1 calculate the profits at the ith MeNB and the jth RN on the fth RB of the time t, respectively. The resource allocation scheme proposed in Proposition 5.1 maximizes the aggregate profit of the network. Note that the profits of the MeNB and the RN are mutually dependent. Reflected in the mathematical formulas, $\lambda_{i}^{f}(t)$ and $k_{1,j}^{*}$ as given in (5.38) and (5.39) are functions of each other.

In the following, one way of jointly calculating $\lambda_{i}^{f}(t)$ and $k_{1,j}^{*}$ is provided.

The values of $\lambda_{i}^{f}(t)$ and $k_{j,1}^{*}$ can be solved from (5.38) and (5.78) using the following procedure.

1. For each $j \in \mathcal{J}_i$ and each $k_{j,1} \in \mathcal{K}_{j,i}$, calculate a $\tilde{\lambda}^f_{i,B}(t)$ as

$$\tilde{\lambda}^f_{i,B,k_{j,1}}(t) = \max_{k_2 \in \mathcal{K}^c_{j,i}} \frac{R^{c,f}_{k_2,j,i}(t)(1 + a^f_{k_{j,1},j,i}(t))}{R_{k_2}(t-1)(1 + a^{c,f}_{k_2,j,i}(t))} - \frac{R^f_{k_{j,1},j,i}(t)}{R_{k_{j,1}}(t-1)}. \quad (5.81)$$

2. Compare $\tilde{\lambda}^f_{i,B,k_{j,1}}(t)$ and $\lambda^f_{i,A}(t)$. If $\lambda^f_{i,A}(t) > \max_{j \in \mathcal{J}_i, k_{j,1} \in \mathcal{K}_{j,i}} \tilde{\lambda}^f_{i,B,k_{j,1}}(t)$, then $\lambda^f_i(t) = \lambda^f_{i,A}(t)$, else proceed to step (3).

3. For each $\tilde{\lambda}^f_{i,B,k_{j,1}}(t)$, calculate $\tilde{k}^*_{j,1,k_{j,1}}$ as

$$\tilde{k}^*_{j,1,k_{j,1}} = \arg \max_{k_1 \in \mathcal{K}_{j,i}} \frac{1}{1 + a^f_{k_1,j,i}(t)} \left(\frac{R^f_{k_1,j,i}(t)}{R_{k_1}(t-1)} - a^f_{k_1,j,i}(t)\tilde{\lambda}^f_{i,B,k_{j,1}}(t) \right). \quad (5.82)$$

Determine the set of feasible solutions as

$$\mathcal{S}_{(\lambda_{i,B,k_{j,1}},k_{j,1})} = \{(\tilde{\lambda}^f_{i,B,k_{j,1}}(t),k_{j,1}) | \tilde{k}^*_{j,1,k_{j,1}} = k_{j,1}\}. \quad (5.83)$$

4. If $\lambda^f_{i,A}(t) > \max_{(\lambda_{i,B,k_{j,1}},k_{j,1}) \in \mathcal{S}_{(\lambda_{i,B,k_{j,1}},k_{j,1})}} \tilde{\lambda}^f_{i,B,k_{j,1}}(t)$, then $\lambda^f_i(t) = \lambda^f_{i,A}(t)$, else proceed to the following.

Define $\tilde{\lambda}^{f*}_{i,B}(t) = \max_{(\lambda_{i,B,k_{j,1}},k_{j,1}) \in \mathcal{S}_{(\lambda_{i,B,k_{j,1}},k_{j,1})}} \tilde{\lambda}^f_{i,B,k_{j,1}}(t)$, calculate $\tilde{\mu}^f_{\bar{j}^*,i}(t)$ as

$$\tilde{\mu}^f_{\bar{j}^*,i}(t) = \frac{1}{1 + a^f_{\tilde{k}^*_{\bar{j}^*,1},\bar{j}^*,i}(t)} \left(\frac{R^f_{\tilde{k}^*_{\bar{j}^*,1},\bar{j}^*,i}(t)}{R_{\tilde{k}^*_{\bar{j}^*,1}}(t-1)} - a^f_{\tilde{k}^*_{\bar{j}^*,1},\bar{j}^*,i}(t)\tilde{\lambda}^{f*}_{i,B}(t) \right), \quad (5.84)$$

where $\tilde{k}^*_{\bar{j}^*,1} = \arg \max_{(\lambda_{i,B,k_{j,1}},k_{j,1}) \in \mathcal{S}_{(\lambda_{i,B,k_{j,1}},k_{j,1})}} \tilde{\lambda}^f_{i,B,k_{j,1}}(t)$.

If $\tilde{\mu}^f_{\bar{j}^*,i}(t) \geq 0$, then $\lambda^f_i(t) = \tilde{\lambda}^{f*}_{i,B}(t)$, else set

$$\tilde{\lambda}^f_{i,B,\tilde{k}^*_{\bar{j}^*,1}}(t) = \max_{k_2 \in \mathcal{K}^c_{\bar{j}^*,i}} \frac{R^{c,f}_{k_2,\bar{j}^*,i}(t)}{R_{k_2}(t-1)(1 + a^{c,f}_{k_2,\bar{j}^*,i}(t))}, \quad (5.85)$$

and go back to implement step (4).

The whole procedure stops whenever $\lambda^f_i(t)$ is found. Based on the calculated $\lambda^f_i(t)$ value, the $k^*_{j,1}$ value and the sets $\mathcal{J}_{i,1}$ and $\mathcal{J}_{i,2}$ can be easily obtained.

Based on the obtained optimal $n^{f*}_{k,j,i}(t)$ and $n^{c,f*}_{k,j,i}(t)$ values, the optimal T_b value for scheduling the direct/access link transmission and the backhaul link transmission can be determined as follows

$$T_b = \frac{\beta}{1 - \beta},\qquad(5.86)$$

where

$$\beta = \frac{\sum_{t=1}^{T_w} \sum_{f=1}^{F} \bar{n}^f(t)}{F \times T_w},\qquad(5.87)$$

is the portion of resources spend in direct/access link transmission averaged over a time window of size T_w subframes, and $\bar{n}^f(t)$ is calculated as

$$\bar{n}^f(t) = \frac{\sum_{i=1}^{N_c} \sum_{j=1}^{N_r} \sum_{k=1}^{N_u} \left(n_{k,j,i}^{f*}(t) + n_{k,j,i}^{c,f*}(t) \right)}{\sum_{i=1}^{N_c} \sum_{j=1}^{N_r} \sum_{k=1}^{N_u} \left(\mathbb{1}(n_{k,j,i}^{f*}(t)) + \mathbb{1}(n_{k,j,i}^{c,f*}(t)) \right)},\qquad(5.88)$$

where $\mathbb{1}(x)$ is an indicator function with $\mathbb{1}(x) = 1$ if $x > 0$, and $\mathbb{1}(x) = 0$ otherwise.

5.4 Performance Results and Discussion

The performance of the radio resource scheduling scheme is simulated in an LTE heterogeneous network with a 19-cell 3-sector three-ring hexagonal cell structure. Four RNs are uniformly deployed in each sector. Simulation setup follows the guidelines for Case 1 described in the 3GPP technical reports [7]. The simulated multipath channel model is chosen to be the extended typical urban (ETU) model. Transmit power of the MeNB is 46 dBm (40 W) and transmit power of the RN is 30 dBm (1 W). The UEs are uniformly distributed in the network with an average of 50 UEs per sector. The UEs are traveling at a speed of 3 km/h. The total bandwidth is 10 MHz with 180 kHz for each frequency resource block (RB). The entire frequency band consists of 50 RBs for data transmission.

In Fig. 5.5, the network throughput achieved by the resource allocation scheme with optimal T_b is compared with those from the schemes with fixed $T_b = 2$ and $T_b = 3$, respectively. The case with $T_b = 2$ corresponds to a scenario where backhaul link and direct/access links equally share transmission time. This emulates the case that after receiving signals from the tth subframe, RNs immediately forward the signals in the following $(t + 1)$th subframe without implementing buffering and scheduling. The case with $T_b = 3$ corresponds to a scenario where backhaul link transmission is enabled every two subframes of direct/access link transmissions. RN in this case has buffering and scheduling capability. The system parameters in the simulation are set to be $\delta = 0$ dB, $\sigma = -15$ dB and $\theta = 1/2$, corresponding to the path-loss based mobile association scheme. The network throughput of the heterogeneous networks is expressed as the relative percentage of the throughput of the homogeneous network. Figure 5.6 shows the optimal T_b value. It can be seen

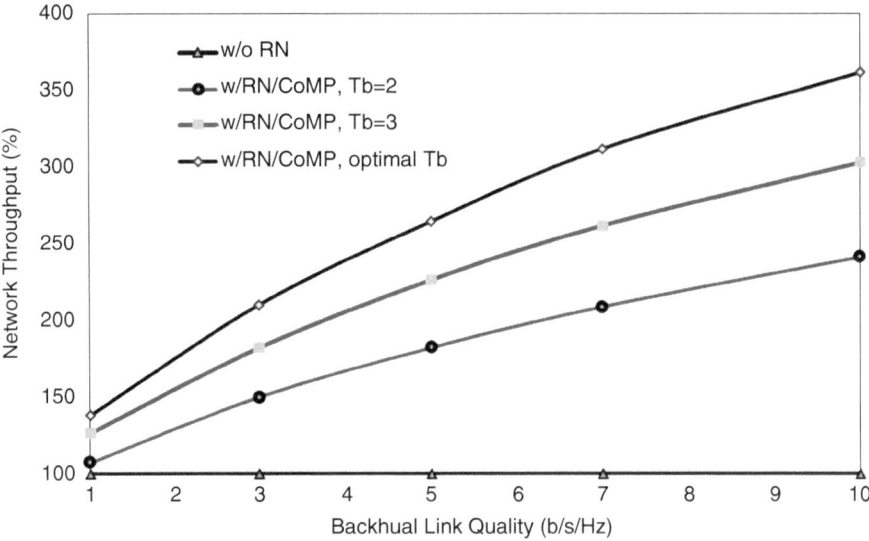

Fig. 5.5 Network throughput comparison for systems with cooperative RN, with non-cooperative RN and without RN

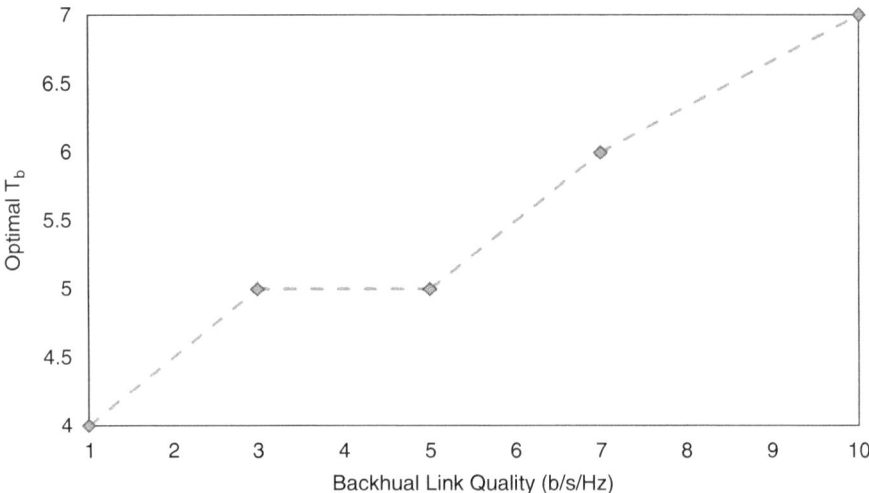

Fig. 5.6 Optimal T_b value for system with CoMP

that with optimal T_b, the resource allocation scheme achieves the best network throughput performance over all backhaul conditions. As backhaul link quality improves, the optimal T_b value increases. When backhaul supports a transmission rate of 10 bit/s/Hz, $T_b = 7$, indicating that backhaul link transmission is activated

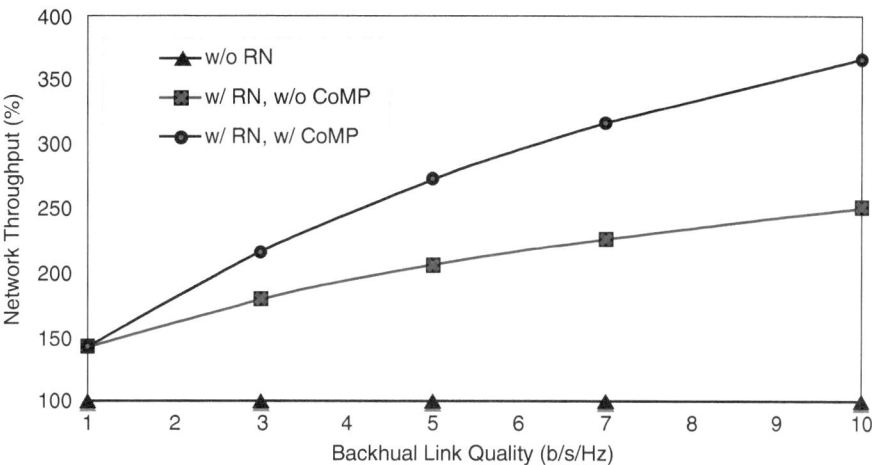

Fig. 5.7 Network throughput comparison for systems with and without CoMP

every 6 subframes of direct/access link transmissions. This result is in consistent with the understanding that as backhaul link quality improves, more resources can be used to support the direct/access link transmission, leading to a better network throughput and a larger time interval between two backhaul transmission instances. Note that in Fig. 5.6, the simulated (backhaul, T_b) points in dashed line are connected for illustration purpose. Other than the simulated (backhaul, T_b) points, the optimal T_b value for the other backhaul qualities should not be directly mapped from the curve.

In Figs. 5.7 and 5.8, performance of the systems with and without CoMP are compared. Asymptotically optimal resource allocation is applied for both CoMP and non-CoMP cases, where asymptotically optimal resource allocation for the non-CoMP systems was proposed in [8]. The system parameters in the simulation are again set to be $\delta = 0$ dB, $\sigma = -15$ dB and $\theta = 1/2$. The throughput advantage due to CoMP can be easily observed from Fig. 5.7. Figure 5.8 demonstrates the average cumulative distribution function (CDF) of the received SINR for the UEs in the macro cell and in the small cell, respectively. The UEs in the small cell include both the R-UEs and the C-UEs. It can be seen that with CoMP, a better CDF behavior can be achieved for the UEs in the small cell while the UEs associated with the MeNB demonstrate a similar CDF behavior for both systems. This observation verifies the performance gain of intra-cell CoMP as observed in Fig. 5.7.

It is understood that network performance would be affected by the adopted association scheme, CoMP UE selection and RN transmit power, i.e., the setting of the δ, σ, θ and P_r values. In Fig. 5.9, the proportional fairness based objective function value under different δ, σ, and P_r values are simulated with fixed $\theta = 1/2$. Setting $\theta = 1/2$ is reasonable to ensure C-UEs a good received signal quality from their respective MeNBs. In Fig. 5.9, systems with different δ values are simulated with fixed $P_r = 30$ dB. It can be observed that for all the simulated mobile

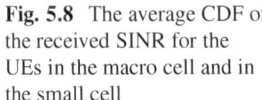

Fig. 5.8 The average CDF of the received SINR for the UEs in the macro cell and in the small cell

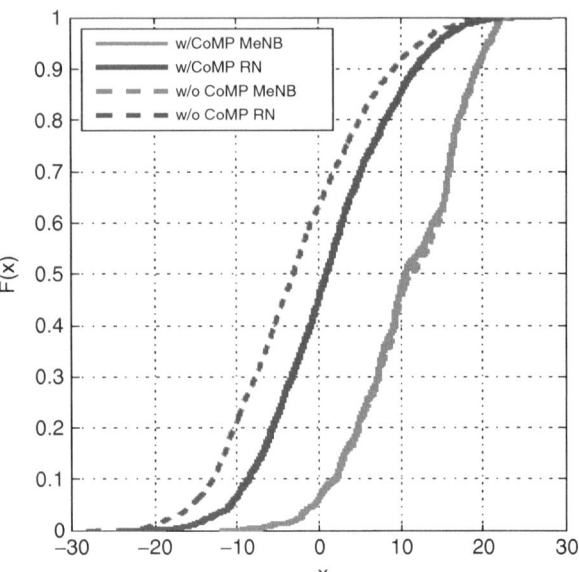

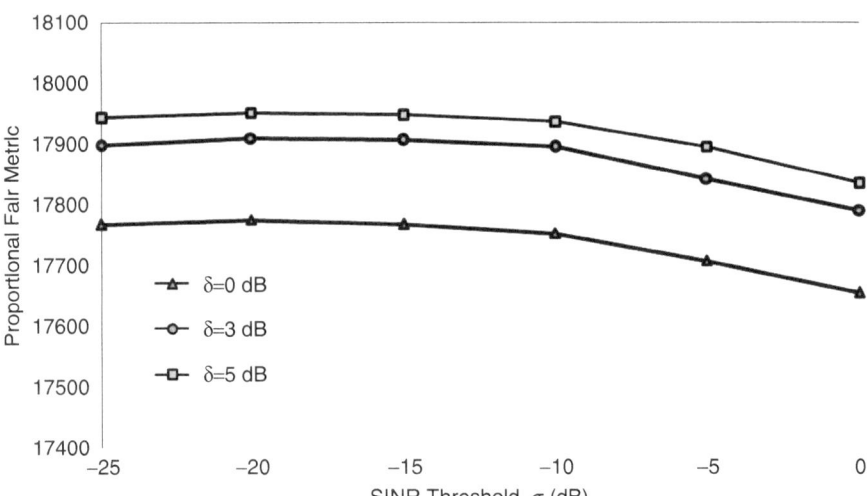

Fig. 5.9 Proportional fairness objective values for systems with different association schemes

association schemes with $\delta = 0, 3, 5$, the optimal SINR threshold for CoMP is $-15\,\text{dB}$. This result indicates that given this same optimal σ value, the number of C-UEs decreases as the value of δ increases. To explain this, note that $\delta = 0\,\text{dB}$ corresponds to path-loss based mobile association and $\delta = 16\,\text{dB}$ corresponds to best-power based mobile association. As the value of δ increases, the coverage range

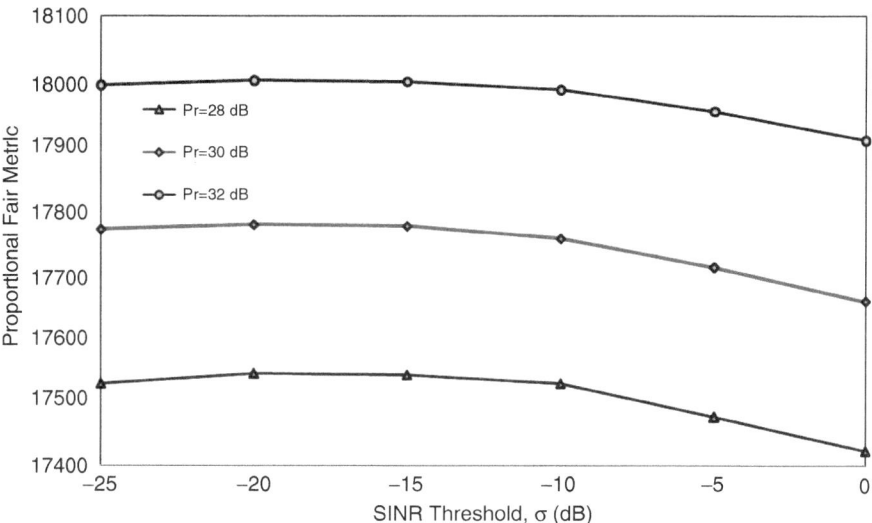

Fig. 5.10 Proportional fairness objective values for systems with different P_r

of RN decreases and the UEs that stay in the RN cell are those with good received signal quality from the RN, leading to a decreasing number of C-UEs. In Fig. 5.10, system simulation is implemented under different RN transmit powers with fixed $\delta = 0\,\mathrm{dB}$. It can been seen that for all P_r cases, the optimal SINR threshold for CoMP is $-15\,\mathrm{dB}$. As P_r increases, the objective function value also increases, which is consistent with intuitive understanding.

5.5 Summary

This chapter presented a resource allocation framework for heterogeneous networks with cooperative RNs and proportional fairness considerations. By applying the gradient-based scheduling scheme and the KKT conditions for optimality, an asymptotically optimal solution is obtained for the framework. The derived asymptotically optimal solution provides a guideline on allocating radio resources among M-UEs, R-UEs and C-UEs. System simulation demonstrates the advantageous performance of the presented resource allocation scheme.

References

1. Y. Xu, R. Q. Hu, "Optimal Intra-cell Cooperation in the Heterogeneous Relay Network," in *Proc. of IEEE Globecom 2012*, Anaheim, CA, Dec. 2012.

2. Y. Xu, R. Q. Hu, Q. C. Li, Y. Qian, "Optimal Intra-cell Cooperation with Precoding in the Wireless Heterogeneous Networks," in *Proc. of IEEE WCNC 2013*, Shanghai, China, April 2013.
3. Q. Li, R. Q. Hu, Y. Qian, and G. Wu, "Intracell Cooperation and Resource Allocation in a Heterogeneous Network With Relays," *IEEE Transactions on Vehicular Technology*, Vol. 62, No. 4, pp. 1770–1783, May 2013.
4. H. Kim and Y. Han, "A proportional fair scheduling for multicarrier transmission systems," *IEEE Commun. Letters*, vol. 9, pp. 210–212, Mar. 2005.
5. R. Agrawal and V. Subramanian, "Optimality of certain channel aware scheduling policies," *Proc. of 2002 Allerton Conference on Communication, Control and Computing*, 2002.
6. A. L. Stolyar, "On the asymptotic optimality of the gradient scheduling algorithm for multiuser throughput allocation," *Operations Research*, vol. 53, no. 1, pp. 12–25, 2005.
7. 3GPP TR36.814, "Further advancements for E-UTRA physical layer aspects," v9.0.0, Mar. 2010.
8. Q. Li, R. Q. Hu, Y. Qian and G. Wu, "A proportional fair radio resource allocation for heterogeneous cellular networks with relays," in *Prof. of IEEE GLOBECOM 2012*, Anaheim, CA, Dec. 2012.

Chapter 6
Conclusion

Global mobile traffic increases 66 times with an annual growth rate of 131 % between 2009 and 2014. On the contrary, the peak data rate from 3G to 4G wireless technology only increases 55 % annually. Clearly there is a huge gap between the capacity growth of new wireless access technologies and the growth of wireless data traffic demand. As wireless channel efficiency is approaching its fundamental limit, future improvements on the wireless capacity are more likely achieved by infrastructure technologies such as node density increase, cooperative and collaborative radio resource management techniques. Moreover, the fast growing data traffic volume and dramatic expansion of network infrastructures will inevitably trigger tremendous escalation of energy consumption in mobile wireless networks, the growing energy consumption becomes one of the major challenges in meeting the cost reduction and green environment targets. To meet those goals, heterogeneous network deployment has emerged as a new trend to enhance the capacity/coverage and to save energy consumption for the next generation wireless networks. A heterogeneous network, or HetNet, is a wireless network containing nodes with different transmission powers and coverage sizes. High power nodes with large coverage areas are deployed in a planned way for blanket coverage of urban, suburban, or rural areas. Low power nodes with small coverage areas aim to complement the high power nodes for coverage extension and throughput enhancement. Furthermore, the infrastructure featuring a high density deployment of low power nodes can also greatly improve energy efficiency compared to the one with a low density deployment of fewer high power nodes, owing to the high path loss exponent in a wireless environment.

As a key technology in 4G-LTE, heterogeneous networks effectively extend the coverage and capacity of wireless networks by deploying multiple small nodes or relay nodes on top of the conventional macro nodes. The deployed small nodes or relay nodes differ in transmission power and processing capabilities, leading to new challenges in mobile association, interference management, and radio resource management. In this book, an in-depth look is provided on the key issues that could affect the performance of heterogeneous networks, and the schemes that can

R.Q. Hu and Y. Qian, *Resource Management for Heterogeneous Networks in LTE Systems*, SpringerBriefs in Electrical and Computer Engineering, DOI 10.1007/978-1-4939-0372-6__6, © The Author(s) 2014

effectively tackle these issues are presented. After an introduction of 3GPP LTE in Chap. 1, the book started in Chap. 2 with a unified HetNet system model in a general LTE system with high transmit power MeNBs and low transmit power SeNBs/RNs. Following that, the concepts on the key radio resource management techniques in HetNets are reviewed, including mobile association, frequency reuse, interference management and cooperative multi-point transmissions. In Chap. 3, a load-balancing based mobile association scheme is presented that optimizes the mobile association by taking account of the traffic load at the MeNBs and SeNBs/RNs, the available resources of the macro and small cells and the network capacity scalability. The load-balancing based mobile association addresses the issues faced in the best-power based and the range-expansion based mobile association schemes and achieves a superior performance. In Chap. 4, an optimization framework is presented for jointly optimizing the frequency subband partition and the transmission power at each sub-band. Fractional frequency reuse is the main component for frequency domain inter-cell interference coordination. It is effective in reducing inter-cell interference and still preserves spectrum efficiency. The challenges in implementing fractional frequency reuse lie in choosing the frequency subband partition between the eNBs and the transmission power level at each subband. In Chap. 5, dynamic radio resource allocation schemes are studied for heterogeneous networks with intra-cell CoMP and relays with in-band backhaul. An optimal dynamic resource allocation framework is presented and a resource allocation strategy that is asymptotically optimal in terms of sum of log scale UE throughput is discussed. The resource allocation scheme gives insights on the optimal radio resource allocation for heterogeneous networks with intra-cell CoMP and relays with in-band backhauls.